American Forage & Grassland Council Annual Conference 2017

Roanoke, Virginia, USA
22 - 24 January 2017

ISBN: 978-1-5108-5016-3

Printed from e-media with permission by:

Curran Associates, Inc.
57 Morehouse Lane
Red Hook, NY 12571

CURRAN ASSOCIATES INC.
proceedings.com

Some format issues inherent in the e-media version may also appear in this print version.

Printed by Curran Associates, Inc. (2018)

American Forage and Grassland Council (AFGC)
PO Box 867
Berea, KY 40403
USA

Phone: 800.944.2342

www.afgc.org

Additional copies of this publication are available from:

Curran Associates, Inc.
57 Morehouse Lane
Red Hook, NY 12571 USA
Phone: 845-758-0400
Fax: 845-758-2633
Email: curran@proceedings.com
Web: www.proceedings.com

TABLE OF CONTENTS

SEEDING YEAR YIELD AND FORAGE NUTRITIVE VALUE OF REDUCED LIGNIN AND CONVENTIONAL ALFALFA SUBJECT TO DIVERSE CUTTING TREATMENTS

A.M. Grev[1], M.S. Wells[2], K.L. Martinson[3], and C.C. Sheaffer[4]

Reduced lignin alfalfa varieties have potential to provide increased management flexibility and to increase the feeding value of alfalfa for livestock animals. The objectives were to determine the seeding year yield and forage nutritive value of a reduced lignin alfalfa variety ('HarvXtra') and conventional varieties when subject to diverse cutting treatments. The experimental design was a randomized complete block with a split plot arrangement of treatments. Whole plots were four cutting treatments, which included 'Standard' (S; 60d + 30d + 30d), 'Standard + Fall' (SF; 60d + 30d + 30d + Fall), 'Standard + Delay' (SD; 60d + 37d + 37d), and 'Delay + Fall' (DF; 67d + 45d + Fall). Sub plots were four alfalfa varieties, which included '54R02', 'DKA43-22RR', 'HarvXtra', and 'WL355RR'. The experiment was seeded in 2015 at two locations in Minnesota. At each harvest date, plots were first hand-sampled to determine forage nutritive value and then mechanically harvested and weighed to determine yield. Cutting treatments SF and DF had greater DM yields than cutting treatment S. At Rochester, yield for conventional alfalfa varieties were higher compared to 'HarvXtra'. At St. Paul, there was no difference in yield among varieties. The S and SF cutting treatments had increased CP, decreased NDF, and increased NDFD concentrations compared to the DF cutting treatment. There were no differences in NDF or CP content among alfalfa varieties. The 'HarvXtra' variety had greater NDFD compared to all other varieties. Cutting treatments and alfalfa varieties affected both yield and forage nutritive values and should be considered when making management decisions.

[1]Graduate Research Assistant, Department of Animal Science; [2]Assistant Professor, Department of Agronomy and Plant Genetics; [3]Associate Professor, Department of Animal Science; [4]Professor, Department of Agronomy and Plant Genetics, University of Minnesota, St. Paul, MN.

GRAZING TERMINOLOGY ACCURATELY DEFINED

John Jennings[1]

Terminology for certain current and popular grazing techniques is poorly defined. This paper offers definitions of popular grazing terms based on practice and outcome.

Mob Grazing (aka **H**igh **S**tock **D**ensity or **HD** grazing) - The practice of combining grazing animals into large groups, or mobs, and grazing a pasture area for short duration.

T.R.A.S.H. Grazing (**T**rample and **R**ecycle **A**ll **S**temmy **H**erbage; also erroneously called Mob grazing) - The practice of concentrating animal stocking density to trample trashy, over-grown, mature, or weedy pasture to facilitate degradation of low quality, high-fiber herbage.

S.M.A.R.T. Grazing (**S**imple **M**anagement **A**nd **R**otation **T**echniques) – This grazing method is a refinement of controlled grazing that emphasizes flexibility in forage management, fencing, and grazing strategy. SMART grazing helps prevent being "boxed-in" with many small permanent paddocks that become obsolete when the last post is driven.

300-Days Grazing – Grazing system developed in Arkansas to reduce hay feeding by planning availability of grazeable forage for each season to extend grazing to at least 300 days per year. Planning forage management practices at least one season or more ahead is a key principle to ensure timely implementation of practices for each part of the grazing season.

[1]Professor and Extension Forage Specialist, University of Arkansas Division of Agriculture. Little Rock, Arkansas 72204.

GRAZING TERMINOLOGY ACCURATELY DEFINED

John Jennings[1]

Terminology for certain current and popular grazing techniques is poorly defined. A grazing technique referred to in one study or presentation may be referred to by a different name by a different speaker. This paper offers definitions of popular grazing terms based on practice and outcome.

Mob Grazing (aka **H**igh **S**tock **D**ensity or **HD** grazing) - the practice of combining grazing animals into large groups, or mobs, and grazing over a specified allocated pasture area for short duration. This practice uses temporary electric fence to allocate forage to a herd at a high stock density of 100,000 lbs/acre animal bodyweight or more. Mob or HD grazing requires short duration grazing periods for each pasture allocation and frequent daily moves to provide adequate forage dry matter and water intake. This practice is typically done to increase grazing efficiency and consumption of high quality forage or to extend grazing of limited forage. It is also an effective tool for managing the forage canopy during transition periods in pastures containing mixed cool and warm season forages. Individual animal performance is reduced due to limited forage selectivity, but forage utilization can be very high. Caution should be followed when Mob-grazing pasture areas containing toxic weeds such as perilla mint to avoid livestock losses.

T.R.A.S.H. Grazing (**T**rample & **R**ecycle **A**ll **S**temmy **H**erbage; also erroneously called Mob grazing) - the practice of concentrating animal stocking density to trample trashy, over-grown, mature, or weedy pasture to facilitate degradation of low quality, high-fiber herbage. The primary goal of TRASH grazing is trampling the fibrous canopy to allow sunlight to reach plant crowns to stimulate forage regrowth. Animals are stocked at a high density per acre for short duration grazing periods using temporary electric fence. In contrast to HD grazing, forage utilization is typically low due to the fibrous nature of the herbage. Because it is used to rejuvenate mature/weedy pastures, it should not be used as a method of providing quality forage for animals with high nutritional requirements. This practice fits best with mature non-lactating animals. TRASH. grazing effectively uses "cows as plows" to bring poor quality pastures back into useful condition. An underlying requirement is the presence of an adequate stand of desirable forage or a soil seedbank of desirable forage species.

[1]Professor and Extension Forage Specialist, University of Arkansas Division of Agriculture. Little Rock, Arkansas 72204.

Simply using TRASH. grazing on areas lacking desirable forage species leads to exposed soil, weed invasion, and high soil erosion potential. Caution should be followed when TRASH.-grazing pasture areas containing toxic weeds such as perilla mint to avoid livestock losses. This method could also be referred to as T.R.A.M.P.L.E. Grazing (**T**rample & **R**ecycle **A**ll **M**ature **P**lant **L**itter **E**fficiently).

S.M.A.R.T. Grazing (**S**imple **M**anagement **A**nd **R**otation **T**echniques) – This grazing method is a refinement of controlled grazing that emphasizes flexibility in forage management, fencing, and grazing strategy to match the forage resource, livestock, and farm goals. SMART grazing makes use of limited permanent electric fences for large pasture subdivisions combined with temporary electric fence to manage daily or weekly grazing allotment for each season as needed (strip-grazing, cell grazing, creep grazing, etc.). The use of temporary electric fence within the large subdivisions created by permanent fence allows convenience for overall field and equipment operations and quick return to a grazing setup. This approach helps prevent being "boxed-in" with many small permanent paddocks that become obsolete when the last post is driven.

300-Days Grazing – The grazing system developed in Arkansas to reduce hay feeding by planning availability of grazeable forage for each season to extend grazing to at least 300 days per year. The emphasis is on using recommended seasonal forage management to extend grazing of existing forages and adding complementary forages to fill gaps as needed. Planning forage management practices at least one season or more ahead is a key principle to ensure timely implementation of practices for each part of the grazing season. The 300-Days Grazing system works equally well with cool-season based forage systems (e.g tall fescue) and warm-season based forage systems (e.g. bermudagrass). Similar strategies are being implemented by many states in the Southeast.

300 DAYS GRAZING PROGRAM - "IT'S ABOUT TIME"

John Jennings, Kenny Simon, Shane Gadberry, Paul Beck, Dirk Philipp, and Don Hubbell[1]

The 300 Days Grazing Program has successfully shown economic benefit for extending the grazing season. However, as with any new educational approach, the program has met opposition from some producers who are convinced that it just won't work. The most common comment from producers regarding why they don't adopt 300-day grazing principles is "I just don't have time to do ...... (fill in the blank)." So a new educational approach is being added to the 300 Days Grazing Program called ***"It's About Time".*** Time required for livestock production is a real issue because most producers are part-time with an off-farm job. Time for farm work is at a premium due to so many other time demands. "It's About Time" documents the amount of time spent harvesting and feeding hay vs. time spent on practices to extend the grazing season. The typical winter hay feeding season in Arkansas is 136 days and surveys indicate most producers spend one hour per day feeding hay for a total of 136 hours. Hay harvest labor was approximately 40 hours per harvest for typical size fields and most producers report making two harvests per year for a total of 80 hours. Fertilization, spraying, and other maintenance for hay production totals an additional 20 hours. The average hay harvest and feeding time requirement is 236 hours or the equivalent of 5.9 40-hour workweeks. By contrast, stockpiling combined with strip-grazing one 40-acre pasture (the equivalent of one hay harvest) for winter grazing only required 26 hours and provides 60 days grazing for a 40-cow herd.

[1]Professor and Extension Forage Specialist, Animal Science Program Associate, Associate Professor-Beef Nutrition, Professor-Forages, Associate Professor-Forages, and LFRS Director, University of Arkansas Division of Agriculture. Little Rock, Arkansas 72204.

ASSESSMENT OF A BERMUDAGRASS CORE COLLECTION FOR ABIOTIC AND BIOTIC STRESS TOLERANCE

William F Anderson, Freddie Cheek, Patricio Munoz, and Alexandra Rucker[1]

Bermudagrass (*Cynodon* sp.) is a valuable forage and hay crop in many parts of the world. Highly productive hybrid bermudagrass cultivars have been developed and are now grown as improved pasture on an estimated two to three million hectares in Southern United States. For further improvements, breeders require information on genetic and phenotypic diversity for the selection of diverse parents that will maximize genetic gains for desirable traits. A bermudagrass core collection was replicated and evaluated in test plots for cold tolerance in the mountains near Blairsville, GA; for shade tolerance, and nitrogen use efficiency in Tifton, GA; and for tolerance to bermudagrass stem maggot (BSM) at multiple locations in Georgia and Florida. For cold tolerance, 31 entries did not survive after the first year. Of the remaining 142 entries, 20 accessions had a higher average yield than Tifton 44. For shade tolerance, all genotypes declined in yield during the second year of shade. Thicker stem genotypes Florakirk, African Star, and breeding lines averaged over 8 Mg/ha equivalent. For nitrogen use efficiency, significant yield reduction at the lower N rate was observed starting at the third harvest in year one with total average yield reductions of 35%, 78%, and 73% in years one to three, respectively. At least eight lines were identified as having high tolerance to BSM and high yields. Germplasm with many useful traits have been identified that should be valuable to forage bermudagrass breeders in the United States.

[1]Research Geneticist, USDA-ARS, Tifton, GA (bill.anderson@ars.usda.gov); Research Technician, USDA-ARS, Tifton, GA; Assistant Professor, Dept. of Agronomy, University of Florida (p.munoz@ufl.edu); Graduate Student, Dept. of Agronomy, University of Florida (alex.m.rucker@ufl.edu).

IMPACT OF FESCUE TOXICOSIS DURING MID TO LATE GESTATION ON FETAL GROWTH.

Susan Duckett, J.L. Britt, M.F. Miller, Jr., and J. G. Andrae[1]

Tall fescue is the primary cool-season grass species utilized by grazing livestock in the southeastern U.S. It typically contains an endophyte that reduces animal performance and profitability; ergot alkaloids produced by the endophyte decrease reproductive rates, milk production, growth rates and weaning weights of livestock. Recent studies by USDA-ARS in Lexington, KY have provided insight into why animals grazing novel or endophyte-free tall fescue often give birth to heavier offspring than animals grazing toxic tall fescue pastures. The objective of this study was to evaluate fetal growth in ewes consuming ergot alkaloids found in endophyte-infected tall fescue at two stages of gestation. Thirty-six Suffolk ewes carrying twins at d 35 of gestation were assigned to one of two treatments: endophyte-free tall fescue seed (E-) or endophyte infected seed (E+) from d 35 – d 85 (2nd trimester) and/or d 86 – d 133 (3rd trimester) of gestation creating four diet combinations: E-E-, E-E+, E+E-, and E+E+. Terminal Caesarean sections were performed at d 133 of gestation. There was no difference in placental or overall fetal weight based on feeding toxic fescue in the 2nd trimester. Ewes receiving E+ seed in the 3rd trimester had 14% lower uterine weights and 23% lower total placentome weights. Overall fetal weight decreased 10% for ewes on E+ treatment compared to E- treatment in the 3rd trimester (*P* = 0.002). Individual organ and muscle weights followed a similar patter. Ergot alkaloid exposure during the 3rd trimester lead to reduced fetal growth; however, impacts in the 2nd semester were minimal or were ameliorated by switching to E- seed in the 3rd trimester.

[1]Clemson University.

COMPARISON OF PRODUCTIVITY AND EFFICIENCY OF GRASS-ONLY AND GRASS-LEGUME BEEF STOCKER GRAZING SYSTEMS IN THE SOUTHERN HIGH PLAINS

L.L. Baxter and C.P. West[1]

The imminent depletion of the Ogallala Aquifer demands innovative alternatives to prevent dramatic losses of income when water levels are insufficient for irrigated row-crop production in the Southern High Plains. With the goal to optimize sustainability and productivity with pasture-based beef stocker systems, a grass-legume (GL) pasture system employing novel forages and grazing techniques was compared to a grass-only (GO) warm-season pasture system. Since protein is often the limiting factor in improving performance of stockers grazing warm-season perennial pastures, higher gains are expected with strategic inclusion of legumes such as alfalfa (Medicago sativa L.). Alfalfa was managed as a protein bank so that a limited water resource was concentrated on a small area to produce a high-quality supplement for poorer-quality base pastures. Inclusion of legumes increased stocker gain achieved per acre in all three years. In addition, the GL system did not require more supplemental irrigation than the GO system. Therefore, it was concluded that the inclusion of legumes increased the water use efficiency of the GL system in all years. The forage-livestock systems created in this research provide a better understanding of the stocker gain-water use nexus and will serve as alternatives for producers transitioning from irrigated row-crops into more resource-sustaining systems.

[1]Graduate Research Assistant, Dept. of Plant and Soil Sciences; Associate Professor and Thornton Distinguished Chair of Plant and Soil Science, Dept. of Plant and Soil Sciences, Texas Tech University, Lubbock, TX 79409.

COMPARISON OF TRADITIONAL AND NOVEL NON-DESTRUCTIVE SAMPLING TECHNQIUES FOR SITE-SPECIFIC ASSESSMENT OF BOTANICAL COMPOSITION IN GRASS-LEGUME PASTURE

L.L. Baxter and C.P. West[1]

To better manage forage productivity, it is important to accurately and objectively monitor legume content in grass-legume pastures. Currently researchers rely on subjective visual assessments or labor-intensive botanical hand separations to determine legume content in mixed pastures. The objective of this study was to determine the predictive accuracy of photographic (ImageJ; Microsoft® PowerPoint, PPT) and in-field, reflectance-based procedures in assessing botanical composition in tall wheatgrass-alfalfa [TW-alfalfa; *Thinopyrum ponticum* (Podp.); *Medicago sativa* L.] and Old World bluestem-alfalfa-yellow sweetclover pastures [OWB-legume; *Bothriochloa bladhii* (Retz) S.T. Blake; *Melilotus officinalis* L.]. None of the techniques were sufficient in determining legume content in TW-alfalfa pasture ($R^2 < 0.50$). Visual and PPT analyses were most successful in OWB-legume pasture ($R^2 > 0.80$), however ImageJ and reflectance-based analyses show potential ($R^2 > 0.50$). The accuracy of non-destructive sampling techniques is promising for mixed pastures when the grass and legume components are distinct hues of green. Although the need for more research is recognized, the ultimate goal is to apply these findings to automated scanners that offer producers rapid measurements of legume cover in mixed pastures.

[1]Graduate Research Assistant, Dept. of Plant and Soil Sciences; Associate Professor and Thornton Distinguished Chair of Plant and Soil Science, Dept. of Plant and Soil Sciences; respectively, of Texas Tech University, Lubbock, TX 79409.

EVALUATION OF WINTER ANNUAL COVER CROPS UNDER MULTIPLE RESIDUE MANAGEMENTS: IMPACTS ON SOIL WATER DEPLETION AND CASH CROP PRODUCTIVITY

L.L. Baxter and C.P. West[1]

Even though the benefits of cover crops for nutrient retention and erosion control are well recognized, adoption has been slow in the Southern High Plains because of concerns that cover crops withdraw soil water to the detriment of the summer crop. A small-plot experiment tested the interacting effects of winter irrigation management and cover-crop species on soil water depletion, productivity of the cover crops, and subsequent productivity of a summer Teff crop. Irrigation regimes were dryland and minimum irrigation (supplement rainfall up to 1 inch/ month). Cover crop species included: rye (cv. Maton II), wheat (cv. TAM204), burr medic (cv. Armadillo), hairy vetch (cv. Vallana), rape-kale (cv. Bayou), and an unplanted fallow. Results from year 1 show that rape-kale produced the most winter biomass followed by rye, despite irrigation regime. The presence of a winter cover did not decrease volumetric water content compared to a winter fallow (monitored at 6 depths up to 3 feet). Although winter irrigation regime did not affect the subsequent Teff harvest, Teff yield was lower in the plots following rape-kale than other cover crops or the winter fallow. Another year of data collection is in progress to corroborate these results. The impact of this project is two-fold: strengthen rural communities by ensuring the persistence of profitable agriculture in the region, and stabilizing the soil surface from excessive wind erosion and desiccation.

[1]Graduate Research Assistant, Dept. of Plant and Soil Sciences; Associate Professor and Thornton Distinguished Chair of Plant and Soil Science, Dept. of Plant and Soil Sciences; respectively, of Texas Tech University, Lubbock, TX 79409.

ARTHROPOD AND SOIL MICROBIAL COMMUNITY SIZE AND COMPOSITION IN NATIVE AND INTRODUCED PASTURES

K.B. Bhandari, C.P. West, S.D. Longing and V. Acosta-Martinez [1]

The Texas High Plains has limited water supply for irrigation to produce agricultural crops and livestock. 'WW-B.Dahl' Old World bluestem [OWB] is a highly productive, drought-tolerant grass in dryland and limited-irrigation conditions. Some varieties of OWB are rich in volatile oils which are suspected of deterring some insects. We characterized the size and structure of the ground-dwelling arthropod and soil microbial communities in different pastures treatments. Visual ratings (1-5 scale) were made for fly densities on cattle grazing OWB alone or OWB+alfalfa pastures. Pastures containing OWB had significantly fewer ground-active arthropods than the alfalfa and native mixed grasses in 2014 and 2015. Number of red imported fire ants and harvester ants was greatest in the native mix and near zero in pastures containing OWB in both years. The cattle fly ratings in 2014 were lower for OWB alone than for OWB+alfalfa, but not in 2015 and in 2016. Pastures did not differ in microbial biomass carbon. OWB-alfalfa had greater microbial biomass nitrogen than OWB, alfalfa and native mix in the surface 2 inches of soil. The presence of OWB apparently provided an unfavorable environment for colonization by fire ants and harvester ants while still hosting desirable insects and soil microorganisms. OWB is a desirable grass in the Southern Plains because of drought tolerance and insect deterrence.

[1] Graduate Research Assistant, Texas Tech University, Lubbock, TX (Krishna.bhandari@ttu.edu); Professor, Texas Tech University, Lubbock, TX (chuck.west@ttu.edu); Assistant Professor, Texas Tech University, Lubbock, TX (scott.longing@ttu.edu); Lead Scientist (Soil Microbiologist and Biochemist), USDA-ARS,

Cropping Systems Research Laboratory, Lubbock, TX (Veronica.Acosta-Martinez@ARS.USDA.GOV).

BREEDING OF ANNUAL RYEGRASS (*LOLIUM MULTIFLORUM*) FOR IMPROVED HEAT STRESS AND DISEASE TOLERANCE

Eric D. Billman, Jesse I. Morrison, and Brian S. Baldwin[1]

Cool-season forage grasses are the backbone of a majority of pasture and hay production systems across the country. However, in the South the ability of these grasses to grow vigorously and persist are limited due to both adverse climate conditions and disease-causing organisms. These abiotic and biotic stress factors pose major risks for the short and long-term production of annual ryegrass (*Lolium multiflorum* L.), which is a staple of most southern forage systems. Of most concern with abiotic stresses are high temperatures at planting. Annual ryegrass grows very poorly in temperatures in excess of 90°F, which are common south of Tennessee until late September or early October. Annual ryegrass is also very susceptible to a disease known as gray leaf spot or 'ryegrass blast' that can decimate annual ryegrass stands if planted in conditions that proliferate disease. Plant breeding can be utilized to improve both annual ryegrass growth under hot conditions and disease resistance. To that end, this research has two objectives: 1) Develop a heat tolerant annual ryegrass variety that can be planted earlier in the growing season by screening populations of Marshall annual ryegrass for the ability to germinate and grow at high temperatures (100°F); and 2) Screen populations of a heat tolerant annual ryegrass in the lab and greenhouse for resistance to gray leaf spot and select individuals that are resistant. Early results indicate significantly increased germination over Marshall at high temperatures, and that improvement through further selection and breeding is viable.

[1]Graduate student; Research Associate III; and Professor of Plant Breeding, Dept. Plant and Soil Science, Mississippi State University.

SUMMER-STOCKPILING PASTURE: A NOVEL SYSTEM FOR EXTENDING THE GRAZING SEASON

Matt Booher[1]

In most parts of Virginia, pasture productivity is consistently inconsistent. More often than not, we have a glut of grass in spring, which is hayed from pasture only to be fed out later in the year, as early as September in dry years. For producers who stockpile fall pasture to extend the grazing season, this is a slap in the face. An unconventional practice can solve this problem for many. Virginia Tech's McCormick Farm is an Agricultural Research & Extension Center located in the Shenandoah Valley. Its 500 acres of fescue-based pasture are managed in a grazing rotation for both research and production purposes. For the past nine years, the farm has been using a *summer stockpiling* system to assist in extending the grazing season for their 230-cow herd. It is a strategy developed by the farm that has allowed them to stockpile excess spring growth in the field without making hay, to provide pasture that is limit-grazed from mid-August to mid-October. While this practice provides an important source of emergency forage, its true value is as a strategic tool to enable other acreage to be stockpiled in fall for winter grazing. Used in this way, the summer-stockpiling system effectively allows for a reduction in hay feeding and related cost savings. The McCormick Farm's records show that the summer stockpiling system has resulted in an average of 272 grazing days/year over the past 9 years, ranging from 250 to 330 days depending on weather. In other words, the farm is now feeding hay for an average of only 85 days each year. Compared to the 150 days they used to feed hay, the summer stockpiling system has resulted in an estimated feed savings of $36/cow per year.

The summer-stockpiling system

Pasture is selected in early spring to be summer-stockpiled and grazing is deferred on it from spring green-up through mid-August. Plants are allowed to mature and set seed without any grazing or mowing. Leafy regrowth accumulates below the canopy and, by August, stems and seed heads dry down and begin to deteriorate. The McCormick Farm summer-stockpiles about 25% of their pasture acres each year, while the remainder is rotationally grazed through spring and summer. While applications of nitrogen do boost spring growth, it has not been found to increase yield of the final stockpile. Strip-grazing of the summer stockpile, with a stocking density of around 60,000 lbs. of live-weight per acre, begins in mid-August. The high stocking density afforded by strip-grazing is critical to stretch the forage supply. No back fence is used and pasture is grazed short before moving animals in order to optimize use of the forage. As the summer stockpile is being grazed, nitrogen is applied to other pastures to begin stockpiling fall growth for grazing in winter. When summer stockpiling has been used on approximately 25% of pasture acres, cattle have been able to strip-graze on it for two months or more in late-summer/early fall, allowing for the *fall* stockpiling of 50% of total pasture acreage elsewhere. This model has consistently extended the grazing season into January or later.

As a supplement to the data collected by the McCormick Farm, the focus of this study was to conduct forage quality testing on summer stockpiled pasture to determine its suitability for various classes of beef cattle. In 2015-16, energy and protein of the sward was evaluated from clipped pasture samples in late summer, along with total ergot alkaloid concentration. Additionally, a fistulated steer was used to compare forage quality and toxicity of grazed samples with those of clipped samples.

While conventional wisdom would suggest summer stockpiled pasture has low nutritional value, our testing shows it is adequate for beef cows at any stage of production (including early-lactation). Growing or finishing animals would require some protein and energy supplementation. Summer stockpiled pasture, with its abundant leafy undergrowth, has averaged about 11% crude protein (CP) and 60% total digestible nutrients (TDN). Additional testing using a fistulated steer indicates that cattle select a diet 2-3 percentage points higher. The toxic alkaloid content of summer stockpiled pasture tested no higher than that of surrounding pasture grazed rotationally. Spring fertilization with nitrogen has not been shown to consistently boost the protein content of the final stockpile.

[1]Crop and Soil Agent, Virginia Cooperative Extension, Verona, VA 24482 (mrbooher@vt.edu).

ADAPTIVE GRAZING FOR HEALTHY SOIL

Greg Brann[1]

Adaptive Grazing is recognizing soil as habitat for a diverse population of plants and soil life. Livestock are used as a tool to promote soil health, so land is more productive, flooding is reduced, and sequesters carbon, while supporting healthier animals and requiring fewer outside inputs. Adaptive grazing focuses on managing strategic disturbance, diversity, living roots, and year-round cover to improve soil functions.

[1]Grazing Soil Health Specialist, Natural Resource Conservation Service, Tennessee.

NITROGEN SOURCE AND APPLICATION EFFECTS ON PASTURE PRODUCTIVITY, LEGUME PERSISTENCE, AND FORAGE NUTRITIVE VALUE

Geoffrey E. Brink[1] and Wayne K. Coblentz[2]

Producers may wish to apply manure to pastures over-seeded with red or white clover, or alfalfa. The effects of applying manure to mixed, cool-season pastures should be carefully considered because the nutritive value benefits of legumes may be more important than increased pasture productivity resulting from manure application. We conducted this study to determine pasture and legume response to manure type and application date. Bedded-pack solid or slurry liquid manure were applied to one-year-old red clover-orchardgrass swards in April, June, or August (solid only) to supply 60 lb N/acre. Plots were grazed at vegetative stage five times (three leaves), or harvested for hay (emerged grass seed head) in spring and then grazed at vegetative stage three times. Plots that received no manure produced 7100 lb DM/acre/year and red clover was found in half of the seeded rows (50% stand) at the end of the grazing season. Applying solid or liquid manure in April increased yield 500 lb DM/acre/year. Applying solid or liquid manure in June or solid manure in August reduced annual yield 400 to 700 lb DM/acre/year. Manure application on any date reduced red clover persistence and forage nutritive value. Because the primary benefit imparted by legumes is improved forage nutritive value, producers should apply manure to pastures having little or no legume.

[1]Research Agronomist, USDA-ARS, U.S. Dairy Forage Research Center, Madison, WI 53706; [2] Research Dairy Scientist, USDA-ARS, Institute for Environmentally Integrated Dairy Management, Marshfield, WI 54449.

FALL STOCKPILING FOR WINTER STRIP-GRAZING: A STRATEGY FOR BUILDING SOIL HEALTH

J.B. Daniel[1]

Stockpiling tall fescue for winter grazing is an old practice applicable to a new generation. Some of the foundational research for this production and management strategy was conducted at Virginia Tech nearly 50 years ago. This practice, based on simple and proven management principles, is even more effective today than it was in the early 1970's when it was first being promoted on farms in this state. With the development of portable electric fence technology, the benefits of stockpiling have increased by the ability to strip-graze the stockpiled forage to increase its utilization and further extend the grazing season. In addition to the production benefits, the stockpiling and strip-grazing technique enhances conservation benefits and supports the principles of building soil health, including keeping the soil covered, minimizing soil disturbance and maximizing living roots. Stockpiling and strip-grazing is a cornerstone practice on grazing operations that maximize grazing days and minimize winter hay feeding. The use of this practice has been revitalized in regions of Virginia over the last 6 years by promoting its production and conservation benefits through on farm grazing demonstrations. These structured farm demonstrations, personalize this management strategy, supported by the testimony of the cooperating farmer at a winter field meeting when most of the neighbors are feeding hay. This cooperative effort between the Virginia Forage and Grassland Council with farmers, Extension and Conservation professionals at the local level, has expanded the use of this technique, as the new generation of grassland farmers adopt this practice on their farms each year.

[1]Forage and Grassland Agronomist, USDA-NRCS, Farmville, VA (j.b.daniel@va.usda.gov).

SOIL HEALTH AND GRAZING MANAGEMENT: PUTTING SCIENCE INTO PRACTICE USING GRAZING DEMONSTRATIONS

Katelyn Cox and J.B. Daniel[1]

The Virginia Forage and Grassland Council (VFGC) is known throughout the state for its high quality winter forage conferences and interactive summer forage field tours. These educational events successfully promote and encourage the improved utilization and management of forages to meet production, economic and conservation goals of producers. VFGC was awarded a Conservation Innovation Grant (CIG) from USDA-NRCS-VA, to demonstrate and document changes in pasture productivity and soil health due to the implementation of improved grazing management and use of forages. To accomplish this objective, the VFGC identified four cooperative farmers across the state who committed to host grazing demonstrations as they implemented changes in the use of forages and grazing management specific to their goals. Each host farms received focused consultation, pasture evaluation, grazing plan development and close follow-up during the implementation period. The farms are in different stages of plan development and practice implementation. This CIG has provided an opportunity for these four farmers to put the science and knowledge learned at winter conferences and summer field days into practice, with the key support and follow up from a designated team of agricultural, conservation and experienced graziers from their region of the state. This project is ongoing and will conclude in 2018 with case study summaries and short videos highlighting the benefits to the farmers, their grazing systems and overall soil health.

[1]Grazing Demonstration Technician, Senior Virginia Tech, Department of Crop and Soil Environmental Sciences, Blacksburg, VA (katec94@vt.edu); Forage and Grassland Agronomist, USDA-NRCS, Farmville, VA (j.b.daniel@va.usda.gov).

NITRATE-NITROGEN CONTENT OF ANNUAL WARM-SEASON FORAGES GRAZED BY HORSES

Michelle L. DeBoer, Craig C. Sheaffer, Scott M. Wells, Marica R. Hathaway, Amanda M. Grev, Devan N. Catalano, and Krishona L. Martinson

While all plants contain some nitrate, excessive levels can occur in forages as a result of stress or an abundance of nitrate in the soil. When nitrate levels are too high, they can overwhelm the digestive system of grazing livestock and result in potentially fatal nitrate toxicity. While toxic levels have yet to be determined in the horse, previous research indicates horses can consume approximately 2% nitrate, or 4,600 ppm nitrate-nitrogen before nitrate toxicosis will be observed. High nitrate levels can be observed in forages including sudangrass, sorghum, pearl millet, oats, orchardgrass, and tall fescue. Therefore, nitrate content is a factor that needs to be considered when utilizing a horse pasture. The objective of this study was to evaluate differences in the nitrate-nitrogen content of annual warm-season forages grazed by horses at two maturities. No differences were observed across forage species; however, nitrate values ranged from 1,049 to 9,202 ppm nitrate-nitrogen on a DM basis. While excessive drought and heat were not observed during the grazing seasons, the research team hypothesizes that mineralization of soil nitrogen is likely the cause of the high nitrate values. Since horses had short-term (< 1 week each month), restricted (4 hours per day) grazing of these forages, no symptoms related to nitrate toxicosis were observed.

POTENTIAL OF FORAGE BRASSICAS FOR USE IN PASTURE-BASED LIVESTOCK SYSTEMS

S. Leanne Dillard[1], Ana I. Roca-Fernández[1,2], Melissa D. Rubano[1], and Kathy J. Soder[1]

Brassicas (*Brassica* spp.) are gaining popularity as high-quality forage for pasture due to production during the summer forage slump and in the fall. Inclusion of brassicas in the diet can be limited by glucosinolates, a class of plant compounds that can reduce weight gain and cause off-flavors in meat and milk. The objective of this study was to determine yield, nutritive value, and glucosinolate content of 3 brassicas [‘Appin’ turnip (*B. rapa*), ‘Barisca’ rapeseed (*B. napus*), ‘Inspiration’ canola (*B. napus*)] and ryegrass [‘KB Supreme’ annual ryegrass (*Lolium multiflorum*)] grown in central PA. Forage yield, nutritional composition, and glucosinolate concentration were determined every 2 weeks. Seasonal forage yield was similar among Brassica varieties (1,023 kg DM/ha) and lowest in ryegrass (242 kg DM/ha). While CP was greater in ryegrass than the brassica varieties (33.1 vs. 29.2%), both degradable and soluble protein fractions were lower in annual ryegrass than all brassicas (76 vs. 84% and 18.7 vs. 14.4%, respectively). While all forages were relatively low in NDF and ADF, ryegrass had greater NDF and ADF than brassicas (35.4% vs. 18.7% and 17.5% vs. 14.1%, respectively). Turnip had the greatest total glucosinolate concentration (27.9 mg/g dry), and rapeseed and canola were 40 and 67% lower, respectively. Annual ryegrass contained no glucosinolates. Including brassicas in a cool-season pasture rotation can increase animal productivity and reduce the need for stored feed during mid-summer and late fall, but, glucosinolate levels must be considered to reduce potential negative impacts on animal health and off-flavors in milk and meat.

[1]Pasture Systems and Watershed Management Research Unit, USDA-ARS, University Park, PA,
[2]Dept. Producción Vegetal, Escuela Politénica Superior, Universidad de Santiago de Compostela, Lugo, España.

VISION, TEAMWORK, AND PROGRAMMATIC EFFORTS SUPPORT SILVOPASTURE ADOPTION IN VIRGINIA

J.H. Fike[1], A.K. Downing[2], J.F. Munsell[3], C.D. Teutsch[1], G.S. Frey[4], J.B. Daniel[5], T.A. Mize[2], K.J. Fisher[2], M.J. Adams[6], and G.J. Pent[1]

Silvopasture, integrated tree-forage-livestock production systems, has limited adoption in the mid-Atlantic due to lack of awareness and understanding. Greater research, outreach efforts and "on-the ground" models of successful silvopastures are needed. We are developing new programming to promote silvopasture as a credible management practice by creating a team from diverse academic and technical service backgrounds. We have conducted training for technical service providers (TSPs) that have increased the understanding and acceptance of these systems. Survey evidence indicates TSPs are supporting silvopasture implementation. New university-owned silvopasture sites have been developed and collaboration with producers is expanding the distribution and scope of sites available for demonstration. Partnership with the Natural Resources Conservation Service has supported development of a demonstration site through a Conservation Innovation Grant and made silvopasture adoption a cost-shared practice. Curriculum for teaching high school students about tree and forage management for silvopastures was developed through a graduate student training program, which resulted in parents seeking more information applying these practices on their farms. Silvopastures are complex systems requiring intensive management, and may not be viewed as acceptable by all TSPs and may not be adaptable by all producers. However, initial results indicate high levels of acceptance as these communities are exposed to training and demonstration.

[1]Virginia Tech, Crop and Soil Environmental Sciences Department (jfike@vt.edu); [2]Virginia Cooperative Extension; [3]Virginia Tech, Forest Resources and Environmental Conservation Department; [4]U.S. Forest Service; [5]Virginia Natural Resources Conservation Service; [4]Virginia Department of Forestry.

THE DUNG DIARIES: HOW DIFFERENT BEETLE GROUPS AFFECT DUNG-MEDIATED GREENHOUSE GAS PRODUCTION

Fallon Fowler, Steve Denning, Shuijin Hu, and Wes Watson[1]

Livestock pose a serious problem to the world's sustainability and health. Cattle alone produce ~11% of the world's total greenhouse gases with nearly a third of that produced by cattle dung. Unused or improperly decomposed dung produces the majority of greenhouses gases so it changes from being beneficial process for soil and plants to an influential driver of global warming. Demand for cattle will likely not waiver: human population will continue growing and so will demand for meat, milk, and arable land. Supporting cost-effective sustainability practices in an expanding livestock sector will become increasingly relevant on a local, if not global, scale. However, there may be a natural solution – dung beetles. Dung beetles are thought to prevent fertilizer waste through the process of mineralization and prevent greenhouse gas production through dung aeration. In summary, dung beetles may reincorporate carbon and nitrogen from greenhouses into soil nutrients instead. My research examines whether different kinds of dung beetle groups (Rollers, Tunnelers, Dwellers), alone or in combination, help to efficiently mitigate dung-produced greenhouse gases (CO_2, CH_4, N_2O) and improve soil health. Preliminary data indicate that: 1) dung beetles decrease CH_4, but increase CO_2 and N_2O levels; 2) dung moisture content is correlated with greenhouse gas reduction; and 3) distinct dung beetle groups similarly affect greenhouse gas fluxes. In the future, I plan to document wild dung beetle communities effect on natural greenhouse gas fluxes so that I may establish how quantity, time to colonization, and species influence a farm's sustainability.

[1]Dept. Entomology and Plant Pathology, North Carolina State University, Raleigh, NC 27695.

DOES INDUSTRIAL HEMP HAVE POTENTIAL AS A FORAGE?

B.M. Goff, C.E. Stringer, A.H. Johns, L.C. Harris, J.P Perry, and D.W. Williams[1]

The recent legalization of industrial hemp (*Cannabis sativa*) in many states has led to many questions among producers about its suitability as a forage. Evidence of its potential is virtually nonexistent in the literature and the goal of this research was to quantify the relationship between yield and forage nutritive value for industrial hemp and kenaf (*Hibiscus cannabinus*). As with most forage crops, CP and IVDMD declined as the season progressed, while yield increased until the initiation of flowering. Harvest of kenaf and industrial hemp grown for fiber must be made shortly after planting to preserve forage nutritive value due to their fibrous nature. While this may preclude their use as a traditional forage, these may have potential as a short-season "emergency" forage. The presence of grain helped buffer the loss in forage nutritive value that occurs with plant maturity. However, grain varieties of industrial hemp were limiting in forage yield. Future research is planned to increase the yield of grain in industrial hemps used as a forage through optimization of its agronomic management.

[1]Assistant Professor, Graduate Research Assistant, Undergraduate Research Assistant, Senior Lab Tech, Field Research Tech, and Professor, Department of Plant and Soil Sciences, University of Kentucky.

GROWTH RATES OF COOL-SEASON MIXED PASTURES DEFOLIATED AT A RANGE OF LEAF AREA INDICES

Thomas C. Griggs, Mark L. Satterfield, Jordan M.A. Koos, Zachary B. Freedman, and Charlene N. Kelly[1]

Pasture growth rate varies with developmental stage, environmental conditions, and timing and severity of defoliation. Relationships of growth rate and total-season herbage production to canopy leaf area index (LAI) across defoliation stages including late maturity are not well-defined for cool-season mixed-species swards in the Appalachian region. We hypothesized that pasture growth rate and total-season forage production will be less for relatively few harvests of late-maturity pasture at LAI>8 than for more-frequent harvests at LAI 3-6. We compared LAI and growth rate of perennial pastures containing cool-season grasses, legumes, and non-legume forbs under defoliation frequencies and residual heights simulating continuous and rotational stocking treatments ranging from a) weekly to 8 or 13 cm; b) periodically from 25-30 cm to 6 or 12 cm; and c) periodically from mature hay stage to 6 or 12 cm. Herbage mass was estimated with a calibrated rising plate meter and LAI was estimated from canopy light interception measurements with a line quantum sensor. Responses of growth rate and total-season herbage production to LAI at defoliation will be presented.

[1]West Virginia University

ECONOMICS OF FORAGE-BASED BEEF PRODUCTION SYSTEMS

Greg Halich

Is extending the grazing season further into winter more profitable than feeding hay? Is rotational grazing more profitable than continuous grazing? Are daily moves in a rotational grazing system more profitable than weekly moves? What is the best stocking rate on a given farm? The answer to all of these questions is "It depends". There is no universally correct answer to any of these questions. It depends on the situation. This presentation provides answers to these questions by showing multiple scenarios where the individual can pick the situation that best describes their farm and current market conditions. Key concepts and topics covered include:

- Cost to Graze: pasture maintenance, pasture rent, and labor
- Hay costs: hay, labor, machinery cost, less potential nutrient value
- Marginal (additional) cost to graze an extra day in the winter
- Cost evaluation of different rotation periods with different cow herd sizes
- Extended season grazing vs. decreased stocking rate tradeoff

Spreadsheet tools will be highlighted so that you can customize your own situations (farm type and market conditions) and answer these questions through your own analysis available at: http://www.uky.edu/Ag/AgEcon/pubs/GrazingHayCostCalc.xlsx

Associate Extension Professor, Agricultural Economics, University of Kentucky
(Greg.Halich@uky.edu).

FIELD EVALUATION OF 42-0-0 FOLIAR FERTILIZER ON IMPROVED BERMUDAGRASS VARIETIES FOR FORAGE PRODUCTION

R. A. Speir[1], C.S. Talton[2], R.C. Waldorf[3], and D.W. Hancock[4]

Foliar fertilizers are used by producers to provide micro and macronutrients to various crops when needed. Recently, Grasshopper Fertilizer Company (Mount Vernon, TX) has marketed a liquid N product to forage producers as an alternative to conventional granular or liquid fertilizers. Three on-farm trials were established in bermudagrass hayfields in Madison, Banks and Elbert Counties in Georgia. Grasshopper's 42-0-0 foliar fertilizer product applied at the labeled rate and their product at an N rate equivalent to UGA recommendations were compared with conventional ammonium nitrate, urea-ammonium nitrate, and urea. Fertilizers were applied and forage harvested at 4-wk intervals. In each location, yields were significantly lower from the Grasshopper's 42-0-0 foliar fertilizer product applied at the labeled rate and no different than the untreated control. We conclude that the use of Grasshopper 42-0-0 at the labeled rate will not provide yields that are comparable to conventional N fertilizer at recommended rates, and that using this foliar fertilizer at recommended rates of N is not cost-effective or practical.

[1]Extension Agent, University of Georgia Cooperative Extension, Madison County, Danielsville, Georgia 30633. [2]Extension Agent, University of Georgia Cooperative Extension, Elbert County, Elberton, Georgia 30635. [3]Extension Agent, University of Georgia Cooperative Extension, Banks County, Homer, Georgia 30547. [4]Associate Professor and Forage Extension Specialist, Department of Crop and Soil Sciences, 3111 Miller Plant Sciences Bldg., Univ. of Georgia, Athens, GA 30602. Corresponding Author's email: dhancock@uga.edu.

EFFECT OF WARM-SEASON ANNUAL GRASSES ON FORAGE-FINISHED BEEF PRODUCTION

D.D. Harmon[1], J.L. Lee[2], D.W. Hancock[1], A.M. Stelzleni[2], R.L. Stewart, Jr[2], J.R. Segers[3], and C.D. Teutsch[4]

A 3-year study was conducted to evaluate forage production, animal performance and carcass merit from forage-finished steers grazing four warm-season annual forages. The grazing trial was conducted in 2014, 2015, and 2016 (70, 63 and 56-d, respectively). One of four forage treatments was assigned to 16 pastures (2 acre) in a randomized complete block design with four replications. Forage treatments were sorghum x sudangrass (SS; *Sorghum bicolor var. bicolor*bicolor var. sudanense*), brown midrib sorghum x sudangrass (BMR), pearl millet [PM; *Pennisetum glaucum* (L.) R.Br.], or pearl millet planted with crabgrass [PMCG; *Digitaria sanguinalis* (L.) Scop.]. Each year, two British-cross beef steers (n = 32; 945±48 lbs) were randomly assigned to one of the 16 pastures for forage finishing after being stratified by weight. Each pasture was subdivided into two, 1-acre paddocks for rotational stocking and put and take stocking was employed to maintain a consistent forage allowance. Shrunk body weight was recorded at initiation, middle, and end of each grazing season to calculate total and average daily gain. Once forage became limiting, steers were harvested, hot carcass weight (HCW) was measured immediately before chilling, and chilled carcasses (24-h) were evaluated for yield grade (YG) and quality grade (QG). Statistical analysis was conducted using PROC MIXED in SAS 9.4 (Cary, N.C.) with main effects of treatment, year, and the interaction. Significance was detected at $\alpha = 0.05$ with tendencies defined as $P < 0.10$ Total seasonal forage yield had a tendency ($P < 0.07$) to be higher for BMR compared to PMCG, with SS and PM as intermediates (6.5, 5.4, 5.9 and 6.0 tons/acre, respectively). The average daily gain of the steers had a tendency ($P < 0.07$) to be higher for BMR and PMCG compared to SS and PM (2.18, 2.14, 1.90 and 1.87 lbs/d, respectively). No treatment differences were detected for carcass variables HCW, DP, KPH, 12-th rib back fat, marbling score, YG and QG ($P > 0.05$). These findings suggest that cattle on BMR, SS, PM, and PMCG perform similarly, giving producers the option to match a forage type based on the availability, cost, and their production system.

[1]Dept. of Crop and Soil Sciences, University of Georgia, Athens, GA 30605
[2]Dept. of Animal and Dairy Sciences, University of Georgia, Athens, GA 30605
[3]Dept. of Animal and Dairy Sciences, University of Georgia, Tifton, GA 31793
[4]Dept. of Crop and Soil Environmental Sciences, Virginia Polytechnic Institute and State University, Blacksburg, VA 24060

SOIL CARBON, BULK DENSITY, AND PHOSPHORUS DISTRIBUTION WITHIN TEN SOUTHERN PIEDMONT GRAZING SYSTEMS

T. J. Hendricks, D.H. Franklin, S. Dahal, D. Hancock, L. Stewart, M. Cabrera, and G. Hawkins[1]

Sufficient and consistent distribution of carbon (C) and phosphorus (P) across the pasture can influence and improve production and sustainability in pasture-based grazing systems. The objective of this study was to determine spatial distribution of bulk density (BD), soil C, and soil P in continuously stocked Southern Piedmont pastures as affected by soil classification and management. Samples were collected from 10 pastures (ranging from 22.7 to 53.9 acres) fertilized with only mineral fertilizer. Grazing areas were classified as either areas of interest (AOIs) where cattle tended to frequent, or "Matrix" defined as the remainder of each pasture. Laboratory analysis included soil bulk density, loss-on-ignition carbon (LOI), permanganate carbon (POXC), and Mehlich-1 P. Soil bulk density was greatest (μ = 1.62 g cm^{-3}) in the 2-4 in. soil layer. Median soil C (as LOI and POXC) were greatest in the 0-2 in. soil layer (Medians = 6.16 percent and 1137 ppm, respectively) and both measures were strongly correlated to soil organic C. Soil P was lowest in the 4-8 in. samples (Mdn = 2.12 ppm), regardless of sampling location. Spatial analyses (through MANOVA) determined that bulk densities were lowest 130-325 ft. from hay, water, or shade. Soil C values were lowest in samples 328 ft. or more from a shade source while soil P was highest within 130 ft. of hay feeding. By managing pasture hay, water, and shade sources, producers may also effectively manage soil compaction, C, and P to improve production and sustainability.

---

[1] Graduate Research Assistant, University of Georgia, Department of Crop and Soil Sciences (thendricks10@gmail.com); Associate Professor, Department of Crop and Soil Sciences, University of Georgia (dfrankln@uga.edu); Graduate Research Assistant, Department of Crop and Soil Sciences, University of Georgia; Associate Professor, Department of Crop and Soil Sciences, University of

Georgia; Associate Professor, Department of Animal and Dairy Science, University of Georgia; Professor, Department of Crop and Soil Sciences, University of Georgia; Assistant Professor, Department of Crop and Soil Sciences, University of Georgia; Respectively.

INFLUENCE OF PLANTING DATE ON LATE FALL DRY MATTER YIELD OF SPRING OAT, WINTER OAT, AND WINTER WHEAT

Kenny Simon, Dirk Philipp, John Jennings, and Robert Rhein[1]

Spring oats are frequently fall-planted in northern parts of the US to provide winter grazing, but suitable planting dates for Arkansas have not been evaluated. 'Jerry' oat, 'Bob' oat, and a VNS winter wheat were planted late-August, mid-September, and mid-October at the University of Arkansas-North Farm in Fayetteville to evaluate dry matter yields, nutritive values, and winter hardiness. At the harvest of all plots in late November, Jerry oat resulted in higher DM yields than the other two forages, regardless of planting date. Jerry oat achieved approximately 2,000 lb DM/acre from the early planting date compared with 1,500 lb DM/acre for Bob oat and less than 1,000 lb DM/acre for winter wheat. DM yields for the September and October planting dates were reduced by 50% and 75%, respectively, compared to the initial planting date. When planted in mid-October, Jerry oat yielded less than 500 lb DM/acre. Values of total digestible nutrients were observed to be between 60-80% regardless of forage and planting dates; crude protein was between 17-30% depending on forage and planting date. Jerry oat was more susceptible to winter injury than Bob oat and winter wheat. Frost injury ratings were well above the other two forages and leaf wilting occurred after a single night freezing temperatures. Jerry oat planted in late August have the potential to provide a significant amount of forage for fall/winter grazing.

[1]Animal Science Program Associate, Associate Professor-Forages, Professor and Extension Forage Specialist, and Animal Science Technician, University of Arkansas Division of Agriculture. Little Rock, Arkansas 72204.

INFLUENCE OF PLANTING DATE ON LATE FALL DRY MATTER YIELD OF SPRING OAT, WINTER OAT, AND WINTER WHEAT

Kenny Simon, Dirk Philipp, John Jennings, and Robert Rhein[1]

Livestock producers often plant winter annual forages intended for fall and winter grazing, but are reluctant to plant before October due to the fear of stand failure caused by hot, dry weather conditions. However, delaying planting until October also delays grazeable forage production until late February or early March. Providing adequate, high quality forage for grazing in late October through December is difficult to accomplish with traditional winter annuals such as wheat, rye, or ryegrass. Spring oat can provide an emergency, high quality forage in fall when other forages are less productive, therefore reducing the dependence on stored or purchased feed. Spring oat is often planted in northern states for a quick fall forage crop. Spring oat has a high risk for winterkill, but the yield potential for fall forage is greater than for wheat and other winter annuals. Arkansas demonstration results show that spring (Jerry) oat planted in early September produced forage for grazing by late October. A replicated research trial was conducted at the University of Arkansas Watershed Research and Education Center (WREC) in Fayetteville to determine influence of planting date on late fall dry matter yield of a spring oat (Jerry), a winter oat (Bob), and a winter wheat (VNS) planted in a tilled seedbed.

The small grains were no-till planted on a well-firmed, tilled seedbed on August 31, September 22, and October 13. Each small grain was planted at 115 lb/acre. NPK fertilizer was applied to each plot immediately after planting based on soil test recommendation for winter annual forage production.

Stand observations were made during the fall growing season. Plots planted on August 31 and September 22 were slower to establish than those planted on October 13. This is likely due to above average temperatures coupled with below average rainfall during the month of September. Plots planted on October 13 established quickly, but had limited fall forage production. All plots were harvested on December 2. Plots planted on August 31 and September 22 produced more dry matter (DM) than those planted on October 13 (Figure 1.) Jerry oat produced significantly more DM yield than Bob oat and VNS wheat for each planting. Bob oat produced significantly more DM yield than VNS wheat for the August 31 planting, but not the September 21 or the October 13 planting. VNS wheat produced limited DM yield regardless of planting date. To sustain grazing animals, at least 1,200 pounds of forage dry matter per acre should be available before turn-in. Therefore, only Jerry oat and Bob oat planted on August 31 produced enough forage DM for fall grazing.

[1]Animal Science Program Associate, Associate Professor-Forages, Professor and Extension Forage Specialist, and Animal Science Technician, University of Arkansas Division of Agriculture. Little Rock, Arkansas 72204.

Forage quality samples were collected at harvest. Forage nutritive quality for the small grains was excellent (Table 1). However, forage quality tended to decline with plant maturity. Percent CP and TDN were lower for the August planting date than for the September 22 and October 13 dates. August-planted Jerry oat was in the late boot to early heading stage at harvest and Bob oat ranged from vegetative to boot stage. However, the forage quality exceeded nutrient requirements of all classifications of livestock. Plots planted on October 13 had the highest percentage of CP and TDN because plants were in the vegetative stage.

Winter injury was also measured at harvest. Jerry oat had significant winter injury regardless of planting date. Winter injury on Bob oat and VNS wheat was minor. However, Bob oat and VNS wheat planted on August 31 had more winter injury than for the September 22 and October 13 planting dates. VNS wheat planted on October showed no signs of winter injury.

Early planted Jerry oat offers high quality forage for grazing from late October through December. However, a producer must plant early to maximize yield potential. Delaying planting from August 31 to September 22 reduced fall yield potential by 50% and and delaying planting from September 22 to October 13 reduced yield potential by another 50%. One potential disadvantage of oat is the lack of cold hardiness. Producers should plan on utilizing oat forage by January 1 due to the high potential for winterkill and cold injury. Wheat had limited DM yield in fall, but has less winterkill and more yield potential in spring.

Fig. 1. December 2nd yield of spring oat, winter oat, and wheat planted on three dates

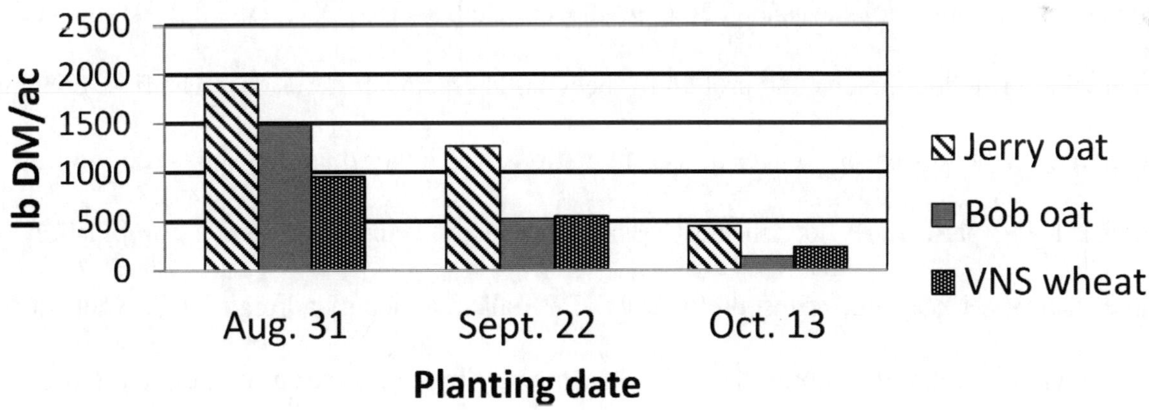

Table 1. Influence of planting date on late fall quality of spring oat, winter oat, and wheat

Planting Date	Jerry oat (spring)		Bob oat (winter)		VNS wheat	
	CP (%)	TDN (%)	CP (%)	TDN (%)	CP (%)	TDN (%)
August 31	17.1	66.8	19.6	72.7	21.6	72.9
September 22	23.3	74.3	29.8	70.5	27.0	80.3
October 13	28.9	74.7	30.9	83.2	30.4	77.7

INTERSEEDING COOL-SEASON FORAGES INTO CORN TO INCREASE YIELD AND QUALITY OF RESIDUE GRAZED IN THE FALL

L. Villalobos[1] and J.E. Brummer[2]

Six forage species and combinations were interseeded into irrigated grain corn to evaluate their potential to improve diets for beef cattle grazing cornstalks during the fall. Species evaluated included annual ryegrass, crimson clover, 'Fridge' winter triticale, a mixture of annual ryegrass plus crimson clover, a brassica combination ('Barkant' turnip, 'Barnapoli' rape, 'Groundhog' radish, and 'Pasja' hybrid), and a combination of winter triticale plus the brassica combination. The cool-season forages were interseeded at the V6 growth stage of corn on June 30, 2014. DM yield (P=0.0013), CP (P=0.0149), aNDF (P=0.0001), and *in-vitro* true digestibility (IVTD, P=0.0027) differed among the interseeded forages. Annual ryegrass and the brassica combination had the highest yields (596 and 790 kg/ha, respectively). The CP content of all species and combinations was higher than the cornstalks (18.3-26.1%vs. 5.2%) and had the potential to provide supplemental protein for beef cattle grazing corn residue. The fiber content of the interseeded cool-season forages and combinations was lower than cornstalks (23.4-44.2% vs. 73.5%), being particularly low in the brassica combination and the brassica combination plus winter triticale. Except for crimson clover (77.7%), all treatments had high IVTD values (89.4-92.1%), values higher than cornstalks (57.7%). Interseeding cool-season forages can improve the overall quality of biomass offered to beef cattle grazing cornstalks during the fall and reduce supplementation costs for producers. This should be combined with strip grazing to maximize utilization of the high quality cool-season forages that can grow during the fall.

[1]Forage Agronomist, Department of Animal Sciences, University of Costa Rica, San José, Costa Rica (luis.villalobosvillalobos@ucr.ac.cr). [2]Forage Specialist, Department of Soil & Crop Sciences, Colorado State University, Fort Collins, CO (Joe. Brummer@colostate.edu)

OVERSEEDING EASTERN GAMAGRASS WITH COOL-SEASON GRASSES OR GRASS-LEGUME MIXTURES

K.M. Bennett[1], M.K. Mullenix[1], J.J. Tucker[2], J.S. Angle[3], R.B. Muntifering[1], and J. Yeager[4]

Overseeding Eastern gamagrass (EG) with cool-season forages may provide an alternative to tall-fescue based forage production systems in the Alabama Black Belt; however, little is known about potential effects on NWSG system sustainability. This study was conducted to determine the effects of overseeding with cool-season forage mixtures and grazing management on forage production, nutritive value, and persistence of Eastern gamagrass. During the cool season, October through April, four overseeding treatments were used in a small plot study: Control (no overseeding), Florida 401 rye (early-maturity), Wrens Abruzzi rye (mid-maturity), and Wrens Abruzzi Rye + AU Red Ace Red Clover. Mob stocking was used to simulate rotational grazing every 28 days and forage samples were clipped at each grazing event to determine forage mass and nutritive value. In the first year cool-season component, Eastern gamagrass overseeded with Florida 401 rye was available earliest to graze, but had less overall forage mass than the Wrens Abruzzi rye or Wrens + Clover mixture. While the majority of the cool-season forage production was later in the season, EG yield response was similar following spring application of N (60 or 120 lb N/acre) in Year 1. A second year of this study is being conducted in 2017.

[1]Department of Animal Sciences, Auburn University, Auburn, AL (kmb0108@auburn.edu), (mullemk@auburn.edu), (muntirb@auburn.edu); [2]Department of Animal and Dairy Science, University of Georgia, Tifton, GA (jjtucker@uga.edu); [3]International Fertilizer Development Center, Muscle Shoals, AL (jangle@ifdc.org); [4]Black Belt Research and Education Center, Auburn University, Marion Junction, AL (yeagejj@auburn.edu)

THE PERSISTENCE OF ORCHARDGRASS (*DACTYLIS GLOMERATA* L.) HAY STANDS

G. B. Jones and B. F. Tracy

Orchardgrass is a globally important long-lived cool-season forage. Recently, producers of orchardgrass hay in the Mid-Atlantic have observed a reduction in the persistence of their swards. Factors related to harvest management, climate, soil, pests/disease, competition, and importantly, their interactions are known to drive the persistence of cool-season grass swards. These factors and interactions will be reviewed based on previous literature and experiments conducted by the authors. Management recommendations and future research needs to improve persistence and productivity will be discussed.

OPPORTUNITIES FOR YIELD MONITORING IN HAY PRODUCTION

Kendall R. Kirk, G. Scott Sell, John Andrae, and Perry Loftis[1]

Yield monitors are a decades-old technology on row crop farms, being commercially available for corn, small grains, and cotton. Many of today's grain combines are equipped with yield monitors as standard equipment. Although work has been done to evaluate and develop yield monitoring systems for hay production, only one system is commercially available and yield monitors are not broadly utilized in hay production. Among the top ten U.S. crops by acreage and the top six U.S. crops by value, hay is third in rank and the only one of these crops where a yield monitor is not widely implemented. The relative absence of yield monitoring for hay production is due to a combination of factors: economies of scale, relative crop management intensity and crop value per unit area, and grower access to complementary technologies.

Being "late-in-the-game" in adoption and development of hay yield monitoring technologies offers a number of benefits and opportunities. Many lessons learned relative to application of yield data in guiding row crop management decisions are transferrable to hay production. Additionally, if hay yield monitors can be developed to be compatible with existing, commercial row crop yield documentation platforms, cost of hay yield monitoring systems will be reduced due to pre-existing sales volumes. Commercialization of a hay yield monitoring system that is suitable for use across all types and manufacturers of balers will be capable of minimizing per unit fixed costs and therefore also minimizing retail prices.

When coupled with a global positioning system, hay yield data can be shown as a map with spatial relationships of the low- and high-yielding areas in a field. By itself, a crop yield monitor generates no reduction in production costs or increases in yield or profit, so determining a generic return on investment for row crop yield monitors has been challenging. However, data collected from the yield monitor can drive management decisions resulting in increased profitability and generally demonstrating payoff periods of one to two years.

A universal application of a hay yield monitor is for direction of crop input rates. A Clemson University study, the beginning of a long-term study with multiple crop inputs, used a hay yield monitor to reveal how profit might be increased in hay production. Prior to onset of this study, this was an intensively managed, irrigated Tifton-85 field with a fixed application of 100 lb N/ac between cuttings. Yield results from this test suggested that if a fixed nitrogen rate was applied across the field, application of 60 lb N/ac would result in $12/ac more profit per cutting when compared to the 100 lb N/ac rate. If yield management zones were utilized for assignment of nitrogen rate by zone, data suggest that profit would have been $4.50/ac greater per cutting than at the 60 lb-N/ac fixed rate. Finally, the Clemson University Directed Prescriptions system suggested that profit would have been $14.50/ac greater per cutting than at the 60 lb-N/ac fixed rate. Environmental stewardship is also improved; when practiced properly, both zone management and directed prescriptions optimize nutrient use in the field, reducing over-application.

The value of any yield monitoring system is dependent on how its information is used to improve management. Return on investment must be evaluated on a case-by-case and farm-by-farm basis. With four cuttings per year, the study discussed here conservatively demonstrated in excess of $50/ac/yr potential benefit from implementation of yield data in prescribing nitrogen rates. If a hay yield monitoring system retailed for $7,500, the data collected here suggests that the system could be paid for in only 150 ac of annual hay production—in nitrogen savings alone.

[1]Clemson University, Clemson, SC.

EFFECTIVENESS OF EXISTING SOIL RHIZOBIA ON CRIMSON CLOVER PERFORMANCE

Robert A. Lane, Shyam Nair[1], Demi Matar, and Brittany McCoy[2]

Cool-season annual legumes are frequently over-seeded into perennial warm-season grass pastures to extend the grazing season, improve forage nutritive value, and provide biologically fixed nitrogen (BNF). Preinoculated clover seed is the industry standard, but research has shown that inoculated seed may be unnecessary in years following successful establishment of the same legume, and that specific strain/cultures/races of *Rhizobium* bacteria are more effective N fixers with specific legume species. We observed how preinoculated and uninoculated 'Dixie' crimson clover performed in a sandy loam soil collected from adjacent sites that had (a) a dense stand of inoculated 'Dixie' crimson clover grown the previous year, (b) inoculated 'Durana' white clover grown for the previous five years, and (c) a control site that had not been planted to a legume but was colonized by naturalized California burr medic. Treatment plants were grown in the greenhouse for 140 days. Inoculation treatment did not affect root or shoot dry matter or N content of crimson clover. Shoot N was higher in plants grown in soil in which crimson clover was grown the previous year. The results indicate that where other legumes have recently grown, existing rhizobial species/races/strains may meet BNF needs of crimson clover for biomass production, but BNF effects on N content of clover (inoculated or not) following inoculated crimson clover appears to be enhanced compared to that following non-crimson legumes.

[1]Professor and Assistant Professor, respectively, Dept. of Agricultural Sciences and Engineering Technology, Sam Houston State University, Huntsville, TX 77341, [2]Former undergraduate students, Department of Agricultural Sciences and Engineering Technology, Sam Houston State University.

STOCKPILING TALL FESUCE FOR WINTER GRAZING IN NORTH MISSISSIPI

R. Lemus, J.A. White, and J.D. Rivera[1]

Feeding hay to livestock is an expensive system in north Mississippi. Stockpiled tall fescue could reduce the need for hay supplementation from mid-November (when perennial warm-season grasses become dormant) to early January (when annual cool-season grasses are available for grazing). The purpose of this demonstration was to evaluate the impact of nitrogen fertilization on stockpiled tall fescue biomass, quality, grazing days and animal performance. Two paddocks were used for this demonstration. One paddock (3.62 acres) contained K-31 tall fescue and the other paddock (2.32 acres) contained 'Texoma' MaxQ II tall fescue. Each paddock was divided in approximately three strips of equal size. Nitrogen treatments of 0, 50 and 100 lb N/ac were randomly applied to the strips within each paddock. Treatments were applied in mid-September of 2014 and 2015 and forage was allowed to accumulate until early December before grazing started. The average grazing length across both seasons for the stockpiled tall fescue was 42 days. 'Texoma' tall fescue had grater ADG and gain per acre compared to K-31 tall fescue. Average nutritive values across N treatments at pre-grazing in 2014 were 14.9% CP, 25.0% ADF, and 49.2% NDF for K-31 and 14.7% CP, 24.0% ADF, and 42.2% NDF for 'Texoma.' In 2015, average nutritive value for K-31 was 19.0% CP, 28.7% ADF and 53.8% NDF while 'Texoma' averaged 18.0% CP, 28.8% ADF and 54.7% NDF.

[1]Extension Forage Specialist, Center for Forage Management & Environmental Stewardship, Mississippi State University (Rocky.Lemus@msstate.edu); Forage Variety Testing Manager, Mississippi Agricultural & Forestry Experiment Station, Mississippi State University (Joshua.White@msstate.edu); Associate Research/Extension Professor, South Mississippi Branch Experiment Station, Mississippi State University (Daniel.Rivera@msstate.edu).

SHADE AFFECTS YIELD, NUTRITIVE CHARACTERISTICS AND ALKALOID CONCENTRATIONS OF FORAGE MIXTURES

Kelly Mercier[1], Chris Teutsch[1], and John Fike[2]

Silvopasture, the intentional incorporation of trees into pasture systems, may moderate the understory microclimate in a way that affects forage productivity. The objective of this experiment was to evaluate the impact of differing shade levels on the yield, nutritive value, and total ergot alkaloid concentration of three cool-season forage mixtures. This study was conducted at Virginia Tech's Southern Piedmont Agricultural Research and Extension Center (Blackstone, VA) with a randomized complete block using four replications. Slatted shade structures were used to create three different shade levels (30, 50, and 70% shade). Three cool-season forage mixtures were evaluated (simple = tall fescue and white clover; intermediate = simple + orchardgrass and red clover; and complex = intermediate + Kentucky bluegrass, birdsfoot trefoil, and alfalfa). Heavy shade reduced yield, but light to moderate shade sometimes increased yield compared to full sun. In most cases shade did not affect total ergot alkaloid concentrations, but they were consistently highest in simple mixtures. Nutritional changes for acid and neutral detergent fibers, crude protein, and total digestible nutrients by harvest indicate forage response to shade varied by season, generally with better forage nutritive characteristics occurring under full sun in the spring and under shade in the summer. Shade influenced yield and most nutritive parameters, while mixture complexity had less of an effect, except for ergot alkaloid concentrations.

[1] Graduate Research Assistant, Virginia Tech's Southern Piedmont Agricultural Research and Extension Center, Blackstone, VA (kmerc094@vt.edu); Associate Professor, Virginia Tech's

Southern Piedmont Agricultural Research and Extension Center, Blackstone, VA (cteutsch@vt.edu); [2] Associate Professor, Virginia Tech, Blacksburg, VA (jfike@vt.edu)

PERFORMANCE EVALUATION OF CATTLE GRAZING DIPLOID AND TETRAPLOID ANNUAL RYEGRASS

S.M. Montgomery, J.I. Morrison and B.S. Baldwin[1]

Annual ryegrass (*Lolium multiflorum* Lam.) has different cytotypes (ploidy levels). It occurs as diploid (2n=2x) or tetraploid (2n=4x). Tetraploids are derived from diploid cultivars by doubling of chromosomes via colchicine treatment. Paradigm suggests polyploid versions of a given plant species have increased seedling vigor, more robust growth and thus a correspondingly greater yield. However, data suggests no difference in yield between diploid and tetraploid varieties, yet there may be a weight gain advantage in cattle grazing tetraploid cytotypes. Therefore, the objectives of this study are to compare diploid and tetraploid varieties of annual ryegrass to evaluate weight gain for cattle grazing the different varieties. Two diploid ryegrass varieties (Marshall and Tam 90) and their corresponding tetraploid analogues (Jumbo and Nelson) were planted in 2.5-acre pastures replicated four times at Mississippi State University Prairie Research Unit, Prairie, MS. Our target planted date was mid-October 2016. Two steers were randomly assigned to each pasture for an 84 day grazing period. Two-day mean average daily gain of the steers was taken pre-, mid- and post-grazing throughout the season to compare weight gain between cytotypes. Other variables that will be measured included herbage mass, dry matter yield (DMY), in vitro dry matter digestibility (IVDMD), relative forage quality (RFQ), visual canopy cover ratings and Brix. Research is currently underway. We hypothesize that diploid varieties will yield better than the tetraploids because they more cold tolerant. Weight gain will be dependent on yield and palatability of each variety and cytotype.

[1]Masters Graduate Student, Dept. of Plant & Soil Sciences, Mississippi State University, MS (smm1006@msstate.edu); Research Associate III, Dept. of Plant & Soil Sciences, Mississippi

State University, MS, (jim46@pss.msstate.edu); Professor, Dept. of Plant & Soil Sciences, Mississippi State University, MS, (bsb2@pss.msstate.edu).

BERMUDAGRASS STEM MAGGOT: DEVASTATING NEW PEST IN SOUTHERN HAY PRODUCTION

Z. Moore

Bermudagrass stem maggot (BSM) or *Atherigona reversura* is a new invasive species that has invaded the Southeastern United States and is crippling the hay production. This species of *Atherigona* is native to Japan, Indonesia, India, and Hawaii. Bermudagrass stem maggot was first noticed in Georgia in 2010. The damage from this insect occurs at the last node of the stem where the leaf emerges. The larva or maggot burrows into the shoot and feeds, and the leaves above the feeding area die. This feeding usually occurs on the top node. The discoloration of the upper leaves causes the field to look like there has been a light frost. If left uncontrolled, up to 80 percent of the tillers in the field can be affected, resulting in significant yield reduction. As of right now, there are very few options for controlling this pest other than the foliar application of pesticides, mainly pyrethroids. Greenhouse studies indicated that fine stem cultivars are more susceptible to BSM. A field study compared spray controlled plots to unsprayed plots in Tifton in 2015 and 2016. For clippings in mid to late summer, BSM reduced yield of Alicia and Russell over 60%. Tifton 85 and Coastcross II also had some yield reduction but not as severe.

Graduate Student, Dept. of Entomology, University of Georgia, (zmoore2374@gmail.com)

ANNUAL FORAGES AS PART OF PROFITABLE BEEF PRODUCTION SYSTEMS

M.K. Mullenix[1]

While perennial forages provide the basis for beef production systems, annual forages can serve as complementary resources for filling in forage production gaps and reducing stored feed needs. Cool-season annuals such as small grains, ryegrass and clovers offer high quality options for grazing and supporting animal performance. Monocultures or mixtures of these species that differ in their individual growth distribution may be used to provide grazeable forage during the autumn and early winter months. Summer annual grasses can provide high yields and nutritive value, but often require improved grazing management to help maintain forage in the vegetative stage of production. Annual forages can be planted into prepared seedbeds or over-seeded into dormant perennial sods as a way to extend grazing on land that may otherwise be unutilized. These systems may also offer several ecosystem services such as erosion control as grazed cover crops, nutrient recovery, and soil quality benefits through above and belowground plant litter contributions.

[1]Assistant Professor/Extension Beef Systems Specialist, Auburn University, Auburn, AL (mullemk@auburn.edu)

MAXIMIZING PASTURE ECONOMIC VALUE

Jim Munsch

On our grass-fed beef operation and for others like it, there are three outcomes that are managed in pursuit of pasture economic value: yield/consumption, quality and cost. These are interdependent with actions taken to change one having an impact on each other.

The starting point is to look at pasture like the agricultural community looks at other crops. That is, know what is possible through benchmarking, setting desired outcomes, and then measuring the actual outcomes. We measure yield, quality and cost with yield and quality measured throughout the season. Desired outcomes on quality must match the class of animal using the pasture.

To know what to do to reach the goal depends on understanding the effect of each decision on the web. That's the general approach that each producer can learn and use. Each farm will be unique in specific actions that should be taken but decision points and actions will be consistent: species of plants and establishment method, fertility plan, plant management through animal impact on the sword, infrastructure, and cost of each input.

Our farm has continual pasture cover with naturally occurring grass species acclimated to our area. We inter-seed legumes (clovers) to improve yield by capturing nitrogen and quality by increasing digestibility. Following soil tests, we selectively out-winter cattle where nutrients are needed. To effectively recycle nutrients, we use moderate stock densities, move daily and have water in each paddock.

[1]Jim Munsch, Owner/Operator Deer Run Farm, Coon Valley, WI 54623 (jmunsch@mwt.net).

EFFECT OF HAY-RAKE TYPE ON EXOGENOUS ASH CONTENT OF FIRST-CUTTING ALFALFA HAY

Abby E. Neu[1]

Ash content of alfalfa (*Medicago sativa* L.) hay typically ranges from 6 to 8% dry matter (DM) with ash contents > 8% indicating hay that is likely contaminated with soil. The objective of this research was to determine the effect of rake-type on ash content of alfalfa hay. Research was conducted in 2015 in Minnesota (loamy sand; Entic Hapludolls), Pennsylvania (silty loam; Typic Hapludalfs) and Wisconsin (silty loam; Typic Agriudolls) on first-cutting alfalfa hay. Treatments involved consolidating two swaths of mowed hay with a wheel rake, side-delivery rake, rotary rake, or a hay merger. To show changes in ash throughout haymaking, samples were taken from standing forage and from cut swaths in each replicate. Post-raking, samples were taken from windrows made by each rake type and from bales (Minnesota and Wisconsin) or haylage (Pennsylvania) made from raked windrows. Samples were analyzed for ash content by igniting samples in a furnace at 600°C. Environmental interaction prevented the combination of rake treatments across locations, so locations were analyzed separately. In all three states, the merger treatment had the least amount of ash, while the wheel rake produced the greatest amount of ash. Farmers looking to reduce ash content of hay should consider using a hay merger when combining swaths.

[1]Research Assistant, University of Minnesota, Department of Animal Science; Extension Educator University of Minnesota Extension.

IMPROVING SOIL HEALTH THROUGH LIVESTOCK AND PASTURE DIVERSITY: A FARM DEMONSTRATION

Jessica Palmer

This project was funded by USDA-NRCs to create a place where livestock producers can learn different management practices and their effects on livestock health, weight gain, and soil and pasture health. The objectives were to 1) Demonstrate options for improving land stewardship and agricultural productivity of rural land protected by a conservation easement and 2) Pull pieces together to determine how mob-grazing has affected farm profitability by discussing animal health (gains per acre, initial and final weights) and farm financial performance. The project had the following methods: 1) Twice-yearly soil sampling and vegetation surveys of four project fields, 2) Time-lapse photography to monitor fields and provide visual information, 3) Livestock weights and farm expenses were recorded. Soil tests showed an increase in organic matter over time for the four fields, and measurements of soil phosphorus, potassium, pH were all within optimal range. Weedy species prevalence decreased and we observed a strong forage base in the fields because of taking a moderate approach to management practices. Our grassland diversity reports also included breeding bird species, and common and rare butterfly and bee species. A positive net profit was realized in 2015. Our monitoring regime was a snapshot of the pastures at one point in time, but the trend appears to be positive for the pastures based on results of our soil sample tests, plant diversity and pollinator surveys. The farm also turned a meager profit just 2 years after beginning this project.

OVERSEEDING EASTERN GAMAGRASS WITH COOL-SEASON GRASSES OR GRASS-LEGUME MIXTURES

K.M. Bennett[1], M.K. Mullenix[1], J.J. Tucker[2], J.S. Angle[3], R.B. Muntifering[1], and J. Yeager[4]

Abstract

Overseeding Eastern gamagrass with cool-season grasses or grass-legume mixtures can extend the grazing season and provide high-quality forage throughout the winter months. The first year of a 2-yr study was conducted at Black Belt Research and Extension Center in Marion Junction, AL to evaluate early- vs. late-maturing cool season forages and effects on persistence of Eastern gamagrass. The cool-season component compared four overseeding treatments: control, early-maturing Florida 401 rye, mid-maturity Wrens Abruzzi rye, and a mixture of Wrens Abruzzi plus 'Red Ace' red clover. During the warm season, May through September, plots were fertilized with either 60 lb nitrogen (N)/ac or 120 lb N/ac. Measurements of pre-graze forage mass, herbage harvested, and herbage accumulated were calculated from pre- and post-graze forage samples. Hand-plucked samples were collected to determine nutritional value. Ground cover estimations and frequency ratings were taken during months in which Eastern gamagrass was breaking dormancy to determine effects of overseeding on persistence of Eastern gamagrass. The same measurements were taken as Eastern gamagrass began to go dormant at the end of the summer. Mob-stocking was used to simulate rotational grazing every 28 days. Year 1 cool-season results indicate that the inclusion of Wrens Abruzzi rye and red clover allowed for higher forage mass and greater digestibility, by all varieties provided high-quality forage for grazing throughout the winter. The percentage of grass presence decreased throughout the cool-season as weeds and clover increased. Year 1 warm-season results indicate that harvest date had an effect on forage mass production. Forage mass peaked in June, then decreased through August, and increased slightly again in September. Nitrogen-fertilization rate did not affect forage mass production. A second year of this study will be conducted.

Introduction: The Black Belt region of Alabama is characterized by heavy clay soils with varying pH extremes. The soil type creates challenges for forage system adaptation in the area. Tall fescue has typically dominated the Black Belt region; however, during the summer months, endophyte-infected tall fescue causes a decrease in animal productivity, specifically in grazing beef cattle. Alternative forage systems could help alleviate the negative effects of endophyte-infected tall fescue on cattle. One alternative may be the use of native warm-season grasses systems (NWSG). NWSG are adapted to the Mid-South region of the US, with characteristics such as drought tolerance, high summer yields, and performance under reduced fertility input.

---

[1]Graduate Research Assistant, Department of Animal Sciences, Auburn University, Auburn, AL; Extension Assistant Professor, Department of Animal Sciences, Auburn University; Professor, Department of Animal Sciences, Auburn University; [2]Assistant Professor, Department of Animal and Dairy Science, University of Georgia, Tifton, GA; [3]President & CEO, International Fertilizer Development Center, Muscle Shoals, AL; [4]Director, Black Belt Research and Extension Center, Marion Junction, AL

Eastern gamagrass (EG; *Tripsacum dactyloides*) is a NWSG with adaptability to heavy soil type like those found in the Black Belt region. It has a wide seasonal growth distribution during the summer months and produces yields that can support high stocking rates relative to other native warm-season grasses. It could be used in place of tall fescue in the summer as an alternative grazing system.

Overseeding Eastern gamagrass with cool-season annual grasses or cool-season annual grass-legume mixtures could extend and provide high-quality grazing throughout the winter months. Following winter overseeding, it is important to determine the impacts of this practice on persistence of Eastern gamagrass and to identify an economical fertilization rate that may help the summer crop recover as it breaks dormancy. Evaluating this alternative forage production system may provide cattle producers with additional forage options within the Black Belt region of Alabama.

Objective: The objective of this study was to determine the effects of overseeding with cool-season forage mixtures and grazing management on forage production, nutritive value, and persistence of Eastern gamagrass in the Blackbelt region of Alabama.

Methods and Methods: A 15-yr old Eastern gamagrass stand located at the Black Belt Research and Extension Center in Marion Junction, AL is currently being used for a two-year small-plot grazing evaluation. Plots were 13-ft x 13-ft and set up in a randomized, complete block design with a factorial arrangement of treatments. Treatments included Eastern gamagrass:

- Control plots (no overseeding)
- Overseeded with 'Florida 401' rye (*Secale cereale*)
- Overseeded with 'Wrens Abruzzi' rye
- Overseeded with 'Wrens Abruzzi' rye +'Red Ace' red clover (*Trifolium pretense*)

Eastern gamagrass was overseeded with the above cool-season mixtures during mid-October as the summer forage slows its production and enters into dormancy. During the early summer months, each plot received either 60 or 120 lb N/acre to determine forage yield and persistence responses of Eastern gamagrass following overseeding. Depending on forage availability, cattle were placed on plots every 28 days using a mob stocking technique (Mullenix et al., 2016) to simulate rotational stocking. Mature beef cattle were used to graze plots to a target stubble height and then removed to mimic Extension recommended grazing strategies.

Forage production measurements were collected by taking two pre-graze and two post-graze samples from each plot every 28 days. Forage dry matter production (lb DM/ac) was estimated from the mass of the pre-graze samples for each grazing event. Forage accumulation was calculated from the difference between post-graze herbage mass from a given grazing event and the subsequent pre-graze herbage mass sampling event. Herbage harvested by cattle was determined by using differences between pre-graze and post-graze sample masses of the same grazing event. Hand-plucked samples were collected before each grazing event for nutritive value analysis [specifically, crude protein (CP) and *in vitro* true digestibility (IVTD)].

Ground cover estimations (% of ground covered by the desirable forage species) and frequency ratings (relative presence and distribution of desirable species within the grazed sward) were conducted before each grazing event in March, April, and May as the Eastern gamagrass broke dormancy. This time period was selected to help determine how Eastern gamagrass forage production recovers following overseeding during the transition from spring to summer. The same estimations were completed in the fall (September) as Eastern gamagrass began to go dormant as a measure of change in persistence following the summer grazing season.

Results and Discussion: Data reported are from the 2016 winter and summer grazing season. Rotational stocking was used to manage cool-season forage during the winter months, with an average rotation of 4 to 6 weeks depending on forage availability and weather conditions. Eastern gamagrass overseeded with Florida 401 rye was available earliest to graze beginning on January 28, and provided two subsequent grazing events on March 1 and March 29. Florida 401 rye is an early-maturing rye that supports harvest and grazing earlier than the later-maturing Wrens Abruzzi rye. Using an early maturing variety potentially allows for more harvests versus the later maturing rye. The first harvest of the Wrens Abruzzi rye and Wrens Abruzzi rye + Clover treatments did not occur until later in the season, beginning on March 8, illustrating the later production of these forage types. While forage mass (lb DM/ac) increased throughout the season for each treatment, total forage production levels for the overseeded treatments were low compared to other studies evaluating overseeding of warm-season forages. In a study evaluating the impact of overseeded cool season forages on bermudagrass, cool-season forage mass was reported as greater than 1,600 lb DM/ac in April (Reis et al., 2001). Control treatment plots were not harvested during the cool-season.

The Florida 401 overseeding treatment had the least pre-graze forage mass (FM), herbage harvested (HH), and herbage accumulation (HA), whereas Wrens Abruzzi had the most, with Wrens + Clover being intermediate (Figure 1). Pre-graze FM and HH increased throughout the season, with March having 445 and 266 lb DM/ac, and April having 1527 and 637 lb DM/ac, respectively. Grass and weed presence was not statistically different among treatments. Clover presence was 6% in the Wrens + Clover treatment and differed from the Florida 401 and Wrens Abruzzi treatments (Table 1). As the season progressed from March into April, percent of grass presence decreased (96% to 77%), and weeds and clover increased (2 to 17%, and 2 to 6%, respectively). Crude protein did not differ among treatments, but *in vitro* true digestibility was lower for Florida 401 than Wrens Abruzzi which was lower than Wrens + Clover (Table 2). However, all forage varieties provided a high quality grazing option during the winter months.

Rotational stocking was used to manage warm-season forage during the summer months, with an average rotation of 4 weeks. Plots were fertilized with either 60 or 120 lb N/acre on May 5. Warm-season harvests and grazing events began on May 25, and continued through September 21. Pre-graze FM, HA, and HH values differed across harvest dates having the greatest values in June, decreasing through August, and then increasing slightly in September (Table 3). After the July grazing event, all plots were cut to a stubble height of 10 inches due to maintain forage in a vegetative state. This may have contributed to the decrease in forage mass in August. Nitrogen fertilizer rate did not affect pre-graze FM, HA, and HH (Table 4). Decreased fertility inputs could allow producers to utilize Eastern gamagrass as a lower-cost grazing system throughout the summer months.

Overall, Year 1 results indicated that overseeding with cool-season annuals could extend the grazing season and provide high quality forage throughout the winter. Reduced fertility inputs during the summer could provide a grazing system with lower costs. A second year of this study will be conducted in 2017 to further determine potential effects on long-term EG stand viability.

Literature Cited:

Mullenix, M.K., Sollenberger, L.E., Wallau, M.O., Blount, A.R., Vendramini, J. and Silveira, M.L., 2016. Herbage accumulation, nutritive value, and persistence responses of rhizoma peanut cultivars and germplasm to grazing management. *Crop Science*, *56*(2), pp.907-915.

Reis, R. A., L. E. Sollenberger, and D. Urbano. "Impact of overseeding cool-season annual forages on spring regrowth of Tifton 85 bermudagrass." In International Grassland Congress, vol. 19, pp. 295-297. 2001.

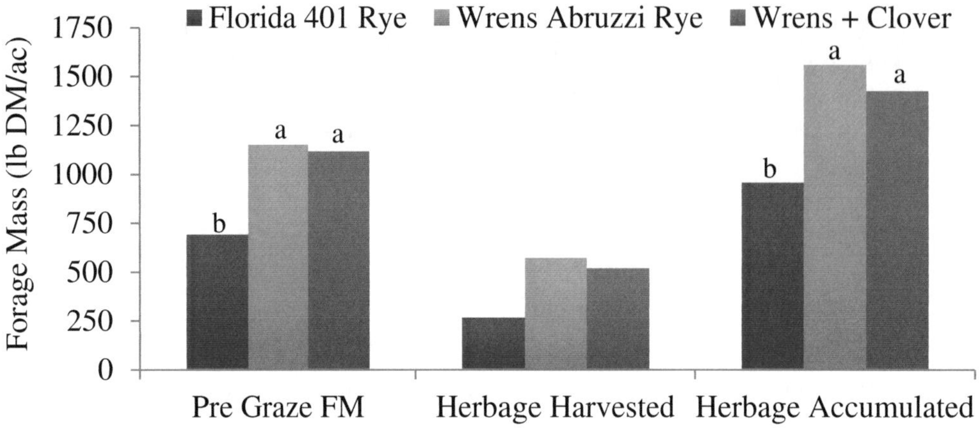

Figure 1. Variety effects on cool-season forage mass production. Means within categories without common superscripts differ $P < 0.05$.

Table 1. Variety effects on cool-season forage botanical composition.

Treatment	% Grass	% Weeds	% Clover
Florida 401 Rye	77	23	0^{b}
Wrens Abruzzi Rye	77	23	0^{b}
Wrens + Clover	77	17	6^{a}

[a]Means within columns without common superscripts differ $P < 0.05$.

Table 2. Variety effects on cool-season forage concentrations of crude protein and *in vitro* true digestibility (% DM basis).

Treatment	% CP	% IVTD
Florida 401	16.3	94.1[b]
Wrens Abruzzi	15.1	95.9[a]
Wrens + Clover	17.2	96.1[a]

[a]Means within columns without common superscripts differ $P < 0.05$.

Table 3. Harvest date effects on forage production (lb DM/acre) of Eastern gamagrass.

Harvest Date	FM (lb DM/ac)	HA (lb DM/ac)	HH (lb DM/ac)
May	537^{be}	---	365^{a}
June	683^{a}	468^{a}	408^{ac}
July	642^{ab}	378^{a}	393^{a}
August	335^{d}	130^{b}	235^{bd}
September	437^{cde}	335^{a}	279^{ad}

[a]Where FM is pre-grazing forage mass, HA = herbage accumulated, and HH = herbage harvested during a grazing event.

[b]Means within columns without common superscripts differ $P < 0.05$.

Table 4. Nitrogen fertilization rate effects on forage production (lb DM/acre) of Eastern gamagrass.

Treatment	FM (lb DM/ac)	HA (lb DM/ac)	HH (lb DM/ac)
120 N	567	343	358
60 N	486	313	314
SE	51	34	42

[a]Where FM is pre-grazing forage mass, HA = herbage accumulated, and HH = herbage harvested during a grazing event.

[b]Means within columns without common superscripts differ $P < 0.05$.

ENHANCED EFFICIENCY NITROGEN FERTILIZER FORMULATION EFFECT ON PASTURE PRODUCTIVITY

K.M. Payne, B.M. Goff, S.R. Smith, J.H. Grove, M.S. Coyne, and J.W. Lehmkuhler[1]

The use of nitrogen (N) fertilizer is generally restricted on mixed species forage systems due to its stimulatory effect on grasses that increases competition with legume species. Reduced legume growth from this competition can compromise the forage nutritive value and prospective yields. The controlled-release nature of several enhanced efficiency (EE) fertilizers holds the potential to improve legume persistence while providing supplemental N required by the grass component. This study evaluated the effect of EE N formulations (ESN, methylene urea, SuperU, and a 75% ESN: 25% urea blend) and untreated urea on forage yield, nutritive value, and legume persistence in a 'Wrangler' bermudagrass and 'Durana' white clover mixture. Nitrogen was applied at four rates (0, 100, 200, and 400 lb N/acre) in two equal applications in each of three growing seasons (2014-2016). During the establishment year, the encroachment of volunteer white clover resulted in only a small decline in the clover population of the highest N rate but became larger among all treatments in the following growing seasons. Seasonal forage yields linearly increased along N rates in all three years, but dry weather conditions in 2015 and 2016 resulted in lower total yield. During the 2014 growing season, ESN did result in higher forage digestibility than other N sources. Continuing research with cool-season mixed pastures and stockpiled forage will allow a comprehensive evaluation of EE N utilization.

[1]Graduate Research Assistant, Dept. of Plant and Soil Sciences, University of Kentucky (kathrynpayne@uky.edu); Assistant Professor, Dept. of Plant and Soil Sciences, University of Kentucky (ben.goff@uky.edu); Extension Forage Specialist, Dept. of Plant and Soil Sciences, University of Kentucky (raysmith1@uky.edu); Director, Research and Education Center-Princeton, University of Kentucky (jgrove@uky.edu); Professor, Dept. of Plant and Soil Sciences, University of Kentucky (mark.coyne@uky.edu); Extension Beef Cattle Specialist, Dept. of Animal and Food Sciences, University of Kentucky (jeff.lehmkuhler@uky.edu).

STOCKPILING COOL SEASON GRASSES IN OHIO

C.D. Penrose and M.A. Landefeld[1]

One way to lower the amount of stored feed needed is to field stockpile forages for fall and winter grazing. Studies have demonstrated that one way to improve the quality and yield of fescue is to apply nitrogen fertilizer (N) when stockpiling is initiated. The most common form of N available in Ohio is urea (46-0-0), but there is a risk of volatilization if adequate rainfall does not occur prior to reacting with the soil. The purpose of these studies is to provide the best options for producers to stockpile cool season grasses to reduce stored feed use. Research over the past three years has been with predominately fescue or orchardgrass pastures. Previous research suggests that initiating stockpiling in early August utilizing 50 lb/acre of N in the form of urea provides increased yields and quality with adequate rainfall. Trials in 2014 and 2015 consisted of urea and urea with a urease inhibitor. The 2014 trial was with fescue and the 2015 trial was with orchardgrass. Both studies were initiated in early August and harvested in early December. In both studies, there was an increase in crude protein content when a urease inhibitor was used. Yields were greater but not statistically significant with the urea and urea with a urease inhibitor in both studies. A replicated trial at three sites is being conducted in 2016 in an attempt to duplicate these results. In addition, ammonium sulfate (21-0-0) was added to the study and an earlier harvest date (November 1) was added to determine if better yields and quality can be achieved with these two new variables.

[1]Ohio State University Extension Educator, McConnelsville, OH (penrose.1@osu.edu); Ohio State University Extension Educator, Woodsfield, OH (landefeld.6@osu.edu).

AUDIO AND TIME-LAPSE IMAGERY ANALYSIS OF LAMB GRAZING BEHAVIOR IN SILVOPASTURE SYSTEMS

Gabriel J. Pent & John H. Fike[1]

Silvopastures – systems where trees are intentionally managed in conjunction with forages – provide distributed shade for grazing livestock. The objective of this study was to determine the grazing behavior and productivity of lambs in silvopasture systems. We compared the growth of lambs in black walnut silvopastures and lambs in honeylocust silvopastures to the growth of lambs in open pastures during two summers. We classified the behavior of a subset of each flock each minute using time-lapse cameras. The same animals were equipped with digital recorders and microphones, and computer software identified grazing behavior from the recordings. Intravaginal temperature sensors logged body temperatures throughout the day. When provided with shade, the lambs spent over 90% of daylight hours within shade boundaries. During the hottest time of day, lambs in the honeylocust silvopastures were 0.5 °F cooler and lambs in the black walnut silvopastures were 0.9 °F cooler than their counterparts in the open pastures. Lambs in the silvopastures spent more time lying down than lambs in the open pastures, yet there was no negative impact on grazing time. Preliminary acoustic analysis indicates that lambs in the silvopastures take as many or more daily bites than lambs in open pastures. The lower level of stress experienced by lambs in the shade positively impacted growth, as all systems produced similar animal gains despite lower forage productivity in some silvopastures.

[1]Gabriel J. Pent, Department of Crop and Soil Environmental Sciences, Blacksburg, VA (gpent@vt.edu); John H. Fike, Department of Crop and Soil Environmental Sciences, Blacksburg, VA (jfike@vt.edu)

DEVELOPING PROTOCOLS FOR CONTAINERIZING HAY FOR EXPORT FROM THE HUMID EASTERN U.S.

K. K. Quick[1], C. D. Teutsch[2], and T.D. Reed[3]

Hay exports from the western U.S. have more than doubled since 2000. The arid conditions found in the west allow for the production of high quality hay for export. However, hay production in this region is highly dependent on irrigation. There is significant interest in developing a hay export market in the humid east. For this to occur, protocols for successfully containerizing hay in the high humidity environment must be developed. The objective of this study was to evaluate the impact of two hay preservatives on the stability of containerized hay. Treatments included 1) propionic acid at baling, 2) propiopnic acid at baling + surface applied propionic acid at container loading, 3) propionic acid at baling + ammoniaization of the container after loading, and 4) no preservatives. Propionic acid was applied at a rate of 5 lb/ton at baling to all treatments except the control. Hay was then stored for 5 weeks before double compressing to a density of 20 lb/ft^2. After compression, bales were placed into containers with treatment 2 receiving a surface application of propionic acid at a rate of 0.0075 lb/bale. After loading hay into the container, treatment 3 was ammoniated at a rate of 3 lb/ton DM. Temperature and relative humidity in the containers were monitored for the 45 day storage period. Hay was sampled prior to compression, at compression, and immediately after opening the containers. Samples were analyzed for neutral detergent fiber, acid detergent fiber, and crude protein. Samples will also be analyzed for mold.

[1]Graduate Research Assistant, CSES, Virginia Tech, Blacksburg, VA (kkquick@vt.edu); [2]Associate Professor, Forage Research and Extension, Virginia Tech's Southern Piedmont AREC, Blackstone, VA (cteutsch@vt.eud); [3]Research Agronomist, Virginia Tech's Southern Piedmont AREC, Blackstone, VA (threed@vt.edu).

SOIL SAMPLING WITHIN HISTORIC LANDSCAPE MANAGEMENT UNITS TO IMPROVE COST EFFECTIVE NUTRIENT MANAGEMENT.

E.B.Rayburn[1]

Adequate soil fertility is required for optimal forage crop and pasture production. However, applying fertilizer beyond crop need is a waste of money and can result in environmental risks. Soil sampling provides information for addressing this issue. However, there may be considerable variability within a field or pasture resulting in "average" soil fertility and "average" fertilizer recommendations that are not biologically or financially optimal. Grid soil sampling is one means to improve the assessment of soil fertility status, and precision fertilizer and lime application can effectively provide for the crop's needs. Where available, this technology can be a cost effective management tool. Where this technology is not available, producers can obtain the same information by dividing fields into smaller management units based on landscape and historic management. Within field management units of 5-acres or less, the technology can provide information that enables the producer to determine if the field is uniform enough to manage as one unit or if the field should be subdivided for lime, phosphorus, and potassium fertilizer applications. With the relatively low cost of soil testing compared to the cost of lime and fertilizer, sampling smaller management units can be an effective tool to reduce input costs while optimizing soil fertility and forage crop and pasture productivity.

[1]Extension Specialist, West Virginia University Extension Service, Morgantown, WV 26506.

PASTURE HERBICIDE STEWARDSHIP: OUR EFFORTS TO REDUCE OFF-TARGET DAMAGE TO SENSITIVE, HIGH VALUE CROPS

G. N. Rhodes Jr. and D. W. McIntosh[1]

Off-target movement of pasture herbicides to high value crops continues to be an issue. Damage can result in lost productivity for growers, expensive fines and/or lawsuits, and negative publicity for the industry. We began a comprehensive educational program in 2011 that stresses the importance of proper stewardship with the use of pasture herbicides. The program was created to reduce the occurrence and impact of off-target damage to sensitive, high value crops, and to make available tools to help with the diagnosis of suspected cases of off-target damage. Funding was obtained via grants from Philip Morris International, Altria Client Services, Dow AgroSciences and DuPont Crop Protection. Four crops (tobacco, cotton, tomato and grape) and five herbicides (2,4-D, dicamba, aminopyralid, aminocyclopyrachlor and picloram) were selected for the development of educational materials and diagnostic tools. These included still images, time lapse videos, and fact sheets that were created and made available when our initial website, *herbicidestewardship.utk.edu*, was launched in 2014. In 2015, the website was completely redesigned in an effort to make it more attractive and user friendly. Additionally, a shorter, less cumbersome address, *herbicidestewardship.com*, was created. Use of the website has steadily increased since its inception. The website has been visited over 9,600 times since launched, and approximately half of these visits were from 2015-2016. Visits came from the United States, China, Japan, Germany, Canada, United Kingdom, India and Brazil.

[1] Professor and Pasture Weed Control Specialist, University of Tennessee, Knoxville, TN (nrhodes@utk.edu); Coordinator of UT Beef and Forage Center, University of Tennessee, Knoxville, TN (dmcintos@utk.edu).

MERGING ECOLOGICAL SITE DESCRIPTIONS AND FORAGE SUITABILITY GROUPS

DWAYNE RICE[1]

Ecological Site Descriptions (ESDs) offer a suitable conceptual platform to capture the ecological processes, potential, and dynamics of Forage Suitability Groups (FSGs). Currently, FSGs are a stand-alone concept but lack correlation with historic plant communities and other alternative converted states of the original range or forest ecological site. Merging FSGs into the ESD model provides valuable soil and plant community information to the public through correlated output products like SSURGO and Web Soil Survey, reduces duplicated documents and description efforts, accesses multiple management interpretations, and captures existing pasture and hay land data into one consistent document format. Incorporating FSGs into ESDs provides customers a better understanding about the sites natural ecological processes, how those processes impact and/or limit production potential, and aid in interpreting and applying management strategies to better mimic these processes.

[1]State Range Management Specialist, Natural Resource Conservation Service, Alexandria, LA 71302.

EFFECT OF FEEDING LEGUMES CONTAINING CONDENSED TANNINS WITH ORCHARDGRASS ON RUMINAL FERMENTATION AND METHANE PRODUCTION IN CONTINUOUS CULTURE

Roca-Fernández, A.I., S.L. Dillard, M.D. Rubano, and K.J. Soder

Developing feeding strategies that reduce livestock methane (CH_4) emissions is gaining interest. Legumes containing condensed tannins (CT) can decrease enteric CH_4 in ruminants, but research is lacking on how CT levels affect ruminal fermentation and CH_4 production. A continuous culture fermenter was used to assess nutrient digestibility, volatile fatty acid (VFA) and CH_4 production, and bacterial protein synthesis of 4 legumes containing varying levels of CT. The legumes were alfalfa (ALF; control), birdsfoot trefoil (BFT; low CT), crown vetch (CV; intermediate CT), and sericea lespedeza (SL; high CT) mixed as 50% orchardgrass and 50% legume. Apparent and true DM and OM digestibility were lower in SL than ALF and BFT, with CV showing an intermediate level. Apparent NDF and ADF digestibility were lowest in SL, without any differences among the other treatments. Total VFA were lower in SL than ALF and BFT, with CV showing an intermediate level. Lower N intake was found in SL than ALF, with BFT and CV showing intermediate levels. Bacterial N efficiency per kg of truly digested DM and OM was lower in SL than BFT, with ALF and CV showing intermediate levels. Total CH_4 production was lower in SL than in the other treatments. No differences were found between ALF, BFT and CV for CH_4 production per g of DM and OM fed or per g of digestible DM and OM fed, with SL showing the lowest CH_4 production per unit of digestible nutrients. In summary, introducing high CT legumes in ruminant diets may be an environmental-friendly feeding strategy for grazing dairy and livestock farmers to reduce CH_4 emissions, but tradeoffs in reduced nutrient digestibility, VFA production, and bacterial N efficiency must be considered.

GRAZING SHEEP ON SOLAR FARMS

Johnny R. Rogers[1]

North Carolina is a leading state in the development of solar electricity generation facilities frequently called "solar farms". Many landowners including farmers are leasing land to solar companies to build solar farms. Maintaining a vegetative cover reduces their environmental impact, but this grass resource must be managed. Mowing is an expensive option, so many solar companies have hired shepherds to graze these facilities or leased the grassland for no charge. Being paid to graze and/or free grass is very attractive. In addition, these facilities are protected by a perimeter fence and panels potentially provide shelter, both of which are beneficial to sheep production. However, managing sheep on a solar farm offers unique challenges. Many shepherds must travel a considerable distance to the solar farms (20+ mile round trip). A water source is not always available, and provisions must be made by digging a well or hauling water. Some solar companies assist shepherds with this expense. The size of solar farms vary (15-1,000 acres) with many in the 30-50 acre range. Stocking rate (live weight/acre/year) must be evaluated to determine the appropriate level because it will be considerably lower than regular pasture systems. Rotational grazing of subdivisions can be achieved with temporary fence, but fencing around solar panels can be difficult and increase labor costs. Portable handling facilities can be used to process the flock. Gathering sheep becomes more difficult as they will move under/between panels when driving them. Grazing sheep on solar farms allows solar energy to be converted to both electricity and animal protein. More research is needed to determine best management practices for this production system.

[1]Dept. of Animal Science, North Carolina State University, Raleigh, NC (jrroger3@ncsu.edu).

THE ROLE OF FORAGE SYSTEMS IN ENVIRONMENTALLY SUSTAINABLE BEEF

C. Alan Rotz and Senorpe Asem-Hiablie[1]

To develop better scientific understanding of the sustainability of beef in the United States, a national assessment is being conducted with support from the Beef Checkoff. This includes a life cycle assessment (LCA) of important environmental, social and economic impact categories of the beef value chain. Forage systems play a major role in the environmental impact of cattle production. Results thus far show farm-gate carbon footprints of representative production systems vary from 16 to 28 kg of carbon dioxide equivalent (CO_2e) per kg of carcass weight (CW) with a mean of 20 kg CO_2e/kg CW. The cow-calf operation is 67 to 77% of this footprint and stocker operations contribute up to 18%. Thus, depending upon whether cattle are backgrounded on pasture or in a feedlot, the grassland-based portion of the system contributes 67 to 85% of the farm-gate carbon footprint of finished beef cattle. Considering post farm-gate sources (harvest, retail, restaurant and consumer), the full carbon footprint is about 45 kg CO_2e per kg consumed beef. Of this total, 58 to 73% can be attributed to emissions from cattle on grassland and the inputs required to maintain them. A similar result is found for environmental impacts such as fossil energy use and total reactive nitrogen loss where 50 to 70% of these farm-gate footprints are attributed to grazing cattle. In the full LCA, grazing animals are also the major contributor to acidification potential and abiotic resource depletion. To make substantial reductions in the environmental impacts of beef, mitigation strategies are needed to reduce greenhouse gas and nitrogen emissions from forage grazing systems. This provides a major challenge because practical strategies for reducing these emissions are essentially unknown.

[1]USDA-ARS, Pasture Systems and Watershed Management Research, University Park, PA (al.rotz@ars.usda.gov and senorpe.asem-hiablie@ars.usda.gov).

STOCKPILING PERENNIAL WARM-SEASON FORAGE GRASSES N EAST-CENTRAL MISSISSIPPI

B. Rushing[1], M. Thornton[2], and R. Lemus[3]

Stockpiling warm-season perennial grass for fall and winter grazing is a management practice that can lower livestock production costs and extend the grazing season. Two seeded bermudagrass and two bahiagrass cultivars were compared in a small plot field trial in east-central Mississippi. Each cultivar had four nitrogen treatments (0, 28, 56, and 84 kg N/ha) with four harvest dates (30, 60, 90, and 120 d after N application) during the fall and winter of 2015 and 2016. Plots were sampled to monitor dry matter (DM), nutritive value [crude protein (CP), acid detergent fiber (ADF), neutral detergent fiber (NDF)], sward height, normalized differential vegetation index (NDVI), and leaf area index (LAI). Total digestible nutrients (TDN) and relative feed quality (RFQ) were calculated based on nutritive values. In 2015, cumulative forage biomass was less than 3000 kg DM ha^{-1} for all cultivars. CP values for all cultivars in 2015 were greater than 110 g kg^{-1} for all harvests and increased with time. Bahiagrass cultivars generally had greater CP than bermudagrass cultivars, with all cultivars remaining above minimum requirements for mature, non-lactating pregnant beef cows (70-80 g kg^{-1}). ADF and NDF increased with time for all entries, with the largest monthly increase occurring in the harvest following a killing frost. TDN and RFQ values suggest energy requirements for beef livestock cannot be met with these standing forages, thus necessitating high energy supplements.

[1]Assistant Research/Extension Professor, Coastal Plain Branch Experiment Station, Newton, MS 39345; [2]Graduate Assistant, Dept. Plant and Soil Sciences, Mississippi State, MS 39762; [3]Extension Forage Specialist, Dept. Plant and Soil Sciences, Mississippi State University, Mississippi State, MS 39762.

NATIVE WARM-SEASON GRASSES: CURRENT RESEARCH IN EAST-CENTRAL MISSISSIPPI

B. Rushing[1] and M. Thornton[2]

Native warm-season grasses (NWSG) are perennial forages that provide quality feed for grazing livestock, and critical nesting and cover habitat for grassland birds and small mammals. Research NWSG pastures is lacking in the Deep South, particularly animal productivity and nutritive value. Two field studies are being conducted to assess NWSG production in east-central Mississippi. The first is a grazing study comparing big bluestem (BBS), mixed NWSG (indiangrass, big bluestem, and little bluestem; MIX), and 'Argentine' bahiagrass (BG). Nutritive value of each grass treatment was monitored from each paddock throughout the study. Crude protein, acid detergent fiber, and neutral detergent fiber indicated that BG had greater quality than the other treatments throughout the 84 day grazing season. Animal productivity [average daily gain (ADG) and total gain per acre] was measured on crossbred steers averaging 541 pounds. Through the first year, ADG for the BBS and MIX treatments was 2.0 lb/head/day or greater for the first 28 d, and remained above 0.95 lb/head/day throughout the study. BG treatments were below 0.98 lb/head/day during the grazing season. The second experiment was a plot trial comparing four harvest intervals (30, 40, 60, and 120 d) and four nitrogen treatments (50, 100, 150, and 200 lbs N/acre/year) in nine mono- and polycultures of NWSG species. Dry matter (DM) yield and nutritive value were measured for each plot at each harvest. N treatments had little effect on cumulative DM yield for most species. Both cultivars of switchgrass ('Alamo' and 'Cave-in-Rock') had greater DM yields compared to all other species.

[1]Assistant Research/Extension Professor, Mississippi State University, Coastal Plain Branch Experiment Station, Newton, MS 39345; [2]Graduate Assistant, Mississippi State University, Dept. Plant and Soil Sciences, Mississippi State, MS 39762.

STOCKER AND 'NELSON' ANNUAL RYEGRASS RESPONSE TO DIFFERENT STOCKING RATES

G. Scaglia[1]

'Nelson' annual ryegrass (*Lolium multiflorum*, Lam.) is a tetraploid cultivar developed for high forage production. It may support greater stocking rates than diploid cultivars like 'Marshall' which has been evaluated in south Louisiana for several tears. Despite the fact that the effect of stocking rates on animal and forage production is site specific, this information is needed for the correct management of the forage resources of a particular area. We used stocker crossbred steers with 550 pounds of average body weight in three consecutive grazing seasons. From previous work using 'Marshall', we knew that a stocking rate (SR) of 1.8 steers/acre was appropriate to get, at the same time, good average daily gains maximizing gain per acre. Four SR were evaluated; 1.5 (SR1), 1.8 (SR2), 2.1 (SR3), and 2.4 (SR4) steers/acre. The number of grazing days were greater for SR1 in year 1 (112.6 days) followed by SR1 and SR2 in year 2 (98 and 94 days, respectively). Due to excessive rainfall during the fall in year 3, grazing days for SR3 and SR4 were the lowest at 41 and 40 days, respectively. Steers' average daily gains were greater for SR1 and SR2 in year 2 (2.77 and 2.85 lb, respectively) and SR1 in year 3 (2.68 lb). Final forage allowance was greatest for SR1 (0.70), followed by SR2 (0.52) while FA in SR3 and SR4 were smaller and similar between them (0.42 and 0.40, respectively). Smaller SR allowed for greater average daily gains and gain/acre. Environmental conditions affected grazing days which in turn affected gains.

[1]Professor, Louisiana State University Agricultural Center Iberia Research Center, Jeanerette LA (gscaglia@agcenter.lsu.edu).

EFFECTS OF CONTINUOUS VERSUS ROTATIONAL EQUINE GRAZING ON PASTURE BOTANICAL COMPOSITION AND YIELD

Shayan Ghajar[1], Bridgett McIntosh[2], and Taylor Duncan[3]

Horses are selective grazers with patterns of herbivory distinct from other large herbivore species. Rotational grazing is recommended for horse owners by Virginia Cooperative Extension, but studies of the effects of rotational grazing on pasture botanical composition and yield in the state have focused on other animals. This study examined the changes in pasture botanical composition and yield under continuous and rotational equine grazing in the first year of a multi-year study. Both systems consisted of 9 acres grazed by 4 thoroughbred geldings each. The continuous pasture was set-stocked, while the rotational system was subdivided into 4 rotational grazing lanes of 1.75 acres each, two vegetative heavy use areas of 0.6 acres each, and a stone dust heavy use area of 0.1 acres. Geldings in the rotational system were rotated to the next paddock when stubble height reached 4 inches, and vegetative heavy use areas were used as supplemental grazing paddocks as needed. Pasture data collection included percent ground cover, desirable grass and legume cover, weed cover, relative abundance of forage and weed species, and dry matter yield (DM). Data will be analyzed to compare ground cover, percent cover of plant functional groups, species composition, and yield between rotational and continuous grazing systems. This study will be continued for two additional years and results will inform the recommendations of equine Extension professionals in Virginia.

[1]Virginia Tech College of Agriculture and Life Sciences, Middleburg Agricultural Research and Extension Center; [2]Equine Extension Specialist, Virginia Tech College of Agriculture and Life Sciences, Middleburg Agricultural Research & Extension Center; [3]Undergraduate Intern, Virginia Tech College of Agriculture & Life Sciences, Middleburg Agricultural Research & Extension Center.

COMPARING ESTABLISHMENT BETWEEN FED AND FROST SEEDED RED CLOVER INTO 'KY-31' FESCUE PASTURE

Paige K. Smart[1], Matthew H. Poore[1], Ben M. Goff[2], Carrie L. Pickworth[1], and Lori Unruh-Snyder[1]

A number of producers feed seed to cattle, by means of a free choice mineral or supplement, in an attempt to easily sow clovers in their pastures through distribution of seed in fecal matter. A total of four studies were conducted two years (2014-2015, 2015-2016) to test the efficacy of this method, termed endozoochorous, on establishing red clover (RC) in pastures. These included: the impact of seed contact with minerals, viability of the seed following passage through the ruminant digestive tract, and competition after passage with existing grasses and fecal pats. The effect of a limestone/inoculant coating was also tested in each study. The mineral contact study confirmed that prolonged exposure to loose minerals reduced the viability of red clover seeds. Seeds that successfully passed through the digestive tract ranged in viability from 5% (coated, C) to 14% (uncoated, U). A small plot study compared C and U RC seeds established in 'KY-31' tall fescue pastures via either frost-seeded (FS) or mixed in a fecal pat (F). Frost seeding resulted in the highest RC seedling densities and experienced no improvements from dragging. Dragging improved the establishment of RC seeds that are present in fecal pats. The field study [treatment groups: F (C and U) and FS (C and U)], determined that endozoochory was not a successful method of RC establishment in 'KY-31' Fescue pastures, whereas frost seeding of either coated or uncoated seed was successful.

[1]College of Agriculture and Life Sciences, North Carolina State University (pmkenned@ncsu.edu, mhpoore@ncsu.edu, and clpickwo@ncsu.edu)
[2] Department of Plant and Soil Sciences, University of Kentucky (ben.goff@uky.edu)

FORAGE MIXTURES IDEAL FOR ORGANIC DAIRY PRODUCTION IN KENTUCKY AND TENNESSEE

S. R. Smith, S. Taylor and K. Lea

Intensive grazing and forage management is required for successful organic dairy operations and little information exists to measure the feasibility of complex forage mixtures grazed intensively. Forage mixtures containing warm and cool season grasses and legumes were evaluated for yield and forage quality in an organic system in Kentucky and Tennessee. The treatments included 1) Red clover, 2) Orchardgrass, 3) Red clover + orchardgrass, 4) Alfalfa + orchardgrass, 5) Red clover + orchardgrass + novel endophyte tall fescue, 6) Red clover + orchardgrass + crabgrass planted in the summer, 7) Red clover + annual ryegrass + crabgrass planted in the summer, 8) Crimson clover + annual ryegrass followed by sorghum-sudangrass + cowpea and 9) Crimson clover + annual ryegrass followed by crabgrass + annual lespedeza. Encroachment of weeds, including Canada thistle (*Cirsium arvense*) and unwanted grasses were a significant issue in all plots, similar to the issues faced by organic dairy producers.

FORAGE QUALITY ON THE MOVE: LONG-TERM STORAGE OF SWITCHGRASS EFFECTS FORAGE QUALITY

S. R. Smith, K. Lea, T. Keene, G. Olson and G. Roberts

While the biomass market presents new opportunities to producers, it also presents unique challenges. Market volatility can force producers to hold onto biomass longer than expected; proper storage of material can protect the value of the material as well as preserve it for an alternative use as stored forage for cattle. Round bales of switchgrass (*Panicum virgatum*) were stored inside, covered and uncovered on the ground or on pallets for two years. Changes in overall mass as well as forage quality parameters were analyzed and suggest that storing bales inside or covering has the ability to significantly reduce mass losses compared to uncovered storage. Forage quality is significantly impacted by both cover and surface (ground or rock) as many nutrients translocate from one part of the bale to another or leach from the bale completely.

FALL STOCKPILED TALL FESCUE YIELD RESPONSE TO NITROGEN FERTILIZER DEPENDS ON SOIL BIOLOGICAL ACTIVITY

Smriti Pehim Limbu, Miguel S. Castillo, Matthew H. Poore, and Alan J. Franzluebbers[1]

Nitrogen (N) fertilizer application to tall fescue during late summer is common to promote herbage accumulation prior to winter grazing. However, little is known of the N supplying capacity of soil under tall fescue pastures. We hypothesized that pastures with low reservoir of biologically active soil N would respond significantly to late summer N fertilizer application and pastures with high reservoir would not respond as much. Twenty pastures with diversity in management were selected across North Carolina and Virginia. Each field had four treatments (0, 40, 80, and 120 lb N/acre as urea, applied in early September 2015) replicated four times in a randomized complete block design. Dry matter accumulation was determined in mid-January 2016. Soil was sampled at 0-4" depth in August 2015 from each of the four replicate blocks of each field. Soil N supplying capacity (i.e. N mineralization) was assessed with inorganic N accumulation during aerobic incubation at 77°F for 24 days, as well as from the flush of CO_2 during 3 days. The flush of CO_2 was highly related to net N mineralization (R^2 = 0.77). Those sites with greatest yield response to N fertilizer had lowest net N mineralization and flush of CO_2. Our hypothesis was accepted – fall-stockpiled yield response to N fertilizer declined with increasingly greater biologically active soil N. Adjusting fertilizer recommendations for fall-stockpiled tall fescue based on biologically active soil N appears appropriate, but we are repeating the study with additional sites.

[1]Graduate student, Forage Agronomist, and Animal Scientist, North Carolina State University, Raleigh, NC (spehiml@ncsu.edu; mscastil@ncsu.edu; mhpoore@ncsu.edu);Research Ecologist, USDA-ARS, Raleigh, NC (alan.franzluebbers@ars.usda.gov).

EVALUATION OF FODDER PRODUCTION SYSTEMS FOR GRAZING DAIRY FARMS

Kathy J. Soder[1], Bradley J. Heins[2], Jim C. Paulson[2],
Hugh Chester-Jones[2], Aimee N. Hafla[1], and Melissa D. Rubano[1]

Sprouted grains have gained interest among grazing dairy farmers due to high grain prices, grain scarcity (in the organic dairy sector) and challenges in producing high-quality forages. Interest has also been spurred by high-profile advertising, as well as farmer reports of improvements in milk yields, cow health and farm profitability. However, there is little data available on feeding sprouted grains in the Northeast or Upper Midwest. We found a 327% increase in fresh weight yield when barley grain was sprouted for 7 days, but there was a 17% loss in dry matter (DM) and a 21% loss in energy. Nutrient digestibility of fodder was not as good as high-quality pasture. Thus, there would likely be a net loss of digestible energy, which could negatively impact animal performance and increase feed costs. When 9 kg (as fed) of fodder replaced 2.7 kg of a corn-based concentrate in a TMR, milk production and fat, and body weight and condition score of lactating cows were not affected. Cows fed fodder had slightly greater milk protein and greater milk urea N (16.5 vs. 13.5), suggesting that cows fed fodder may utilize protein less efficiently. Income over feed costs (IOFC) favored not feeding fodder except when grain prices increased by 50% over those used in the study ($11.77/bushel), giving fodder a slight edge ($0.44/cow/day). Because initial investment in the fodder system was not included in the IOFC analysis, actual fodder production costs would be higher. Each farm must put pencil to paper to determine if implementing fodder in feeding management is a good idea.

[1]Research Animal Scientists, USDA-ARS, University Park, PA (Kathy.Soder@ars.usda.gov; Aimee.Hafla@gmail.com; melissa@rreng.net); [2]Associate Professor (hein0106@umn.edu), Professor (chest001@umn.edu), Extension Educator (jcp@umn.edu) Dept. of Animal Science, University of Minnesota.

RESIDUAL HERBICIDE EFFECT ON INTERSEEDED ANNUAL CLOVERS IN SOUTHERN FORAGE SYSTEMS

J. J. Tucker[1], M.K. Mullenix[2], S.F. Enloe[3], and P.L. Burch[4]

The use of herbicides in permanent pasture for weed control has been an excellent option for producers in the Southeast. In the Deep South however, producers have expressed concern when over-seeding legume species into warm-season perennial sods post long-residual herbicide application. Few studies have evaluated the effect of herbicides on clovers, and to date none of these have evaluated annual clovers interseeded into bermudagrass sods. The objective of this study was to evaluate residual herbicide effects on crimson clover over-seeded into bermudagrass sod of two commonly used pasture herbicides, with and without residual weed control, and a non-treated control for comparison. Herbicides used were aminopyralid + 2,4-D (1.5 pt/a) and triclopyr + fluroxypyr + 2,4-D (1.5 and 2 pt/a). These treatments were applied in late spring (May '14) and early summer (June '14), and late spring in multiple or single applications across two years (yr) at two locations in Alabama. Data collected over the 2-yr evaluation indicate clover % cover, DM yield, and height were great enough in all plot areas to be considered successfully established clover (≥ 30%) when planted 5.5 months post herbicide application. Regression analysis resulted in a strong correlation for clover % cover ($R^2 = 0.92$) across all treatments, years, and locations, which suggests that as clover cover increases, grass and other species present decreases. These data indicate that annual clover establishment can be successful in over-seeded bermudagrass systems when planted in the season following treatment with these herbicides.

[1]Assistant Professor, Animal and Dairy Science, University of Georgia; [2]Assistant Professor, Animal Science, Auburn University; [3]Associate Professor, Agronomy, University of Florida [4] Field Scientist, Dow AgroSciences.

RESIDUAL EFFECT OF BROADLEAF PASTURE HERBICIDES ON VOLUNTEER WHITE CLOVER STANDS

E.K. Twidwell and R.E. Strahan[1]

White clover (*Trifolium repens* L.) is a very important component of many pastures in the southeast region of the USA. In Louisiana, white clover is classified as either a short-lived perennial or a reseeding annual. It is normally a reliable source of grazing from March through June. A management practice that could potentially have a negative impact on white clover stands is the application of broadleaf herbicides to control broadleaf weeds during the summer months. Several commonly-used pasture herbicides contain active ingredients that remain in the soil for many months after application and may hinder germination and growth of white clover. In this study, six broadleaf herbicide treatments were applied on May 22, 2015 to an established mixed stand of bahiagrass (*Paspalum notatum* Flugge) and white clover. About ten months after herbicide application, GrazonNext HL applied at 1.5 pints/acre and Chaparral applied at 2.5 ounces/acre had the most negative impact on the proportion of white clover in the mixture. The herbicide treatments which had the least negative effect on white clover populations were picloram, 2,4-D and Weedmaster. GrazonNext HL applied at a rate of 1.2 pints/acre had a moderate effect on the white clover population. Results of this study suggest that the application of GrazonNext HL (1.5 pints/acre) and Chaparral herbicides for summer broadleaf weed control should be avoided if white clover is a component of a pasture.

[1]Edward Twidwell (etwidwell@agcenter.lsu.edu), Professor, and Ronald Strahan (rstrahan@agcenter.lsu.edu), Associate Professor, LSU Agricultural Center, Baton Rouge, LA 70803.

ASSESSING PLANT AND SOIL RESPONSES TO STOCKPILING NATIVE WARM SEASON GRASSES

Neal Wepking, Renata Nave-Oakes, Zach Mcfarlane, Neal Eash, and Travis Mulliniks

Prior research shows that stockpiling is a profitable tool for cow-calf producers managing tall fescue (TF) pastures. However, there is limited research on stockpiling native warm-season grasses (NWSG) such as switchgrass and big bluestem/indiangrass blend. Since NWSG pastures require a fall rest period to build energy reserves, a degree of stockpiling normally occurs during this period. In 2015-2016, forage and soil variables were measured in stockpiled paddocks during the winter grazing period (January-April). Nutritional loss in paddocks occurred during fall, but no significant degradation occurred beyond January. During grazing, forage quality variables (CP and NDFD) were higher for TF compared to NWSG paddocks; however the herbage mass produced was higher for NWSG. Forage disappearance was higher in TF (6.07 kg/ha/day) but did not differ in NWSG paddocks. Labile soil pools indicated elevated carbon and nitrogen pools under TF, and similar monthly variation across all paddocks. The loss of nutritive value may be compensated by the additional herbage mass produced by NWSG pastures, indicating that NWSG is potentially economical for supplementary winter forage.

IMPACT OF SORGHUM-SUDANGRASS HYBRID ON ANIMAL PERFORMANCE UNDER MISSISSIPPI CONDITIONS

J.A. White, R. Lemus, and J.D. Rivera[1]

During the summer, cattle pasture in Mississippi usually includes the warm-season grasses bermudagrass, bahiagrass, or dallisgrass. These pastures do well in Mississippi, but can become limited in July during hot dry weather. In contrast, sorghum x sudangrass hybrids can supply vegetative forage during the hottest summer months under limited rainfall, as well as a convenient crop to rotate with annual ryegrass in a prepared seedbed. Two paddocks making up 3.58 acres were planted in June of 2015 and 2016 using 'Green grazer V', a sorghum sudangrass hybrid, at the Leveck Animal Research Center in Starkville, MS. In 2015, ten Brangus crosses averaging 560 lbs initiated grazing on June 23rd and grazed the two paddocks for 36 days. In 2016, sixteen Brangus crosses averaging 647 lbs initiated grazing on June 30th and grazed for 18 days according to available forage. Average daily gain in 2015 was 0.96 lbs and gain per acre was 98 lbs. The 2016 grazing period was considerably less with only 0.63 lbs of average daily gain and 51 lbs of gain per acre. Crude protein was higher in 2016 compared to 2015 (17% vs. 12%) and NDF was lower in 2016 compared to 2015 compared to 2015 (55% vs. 60%) due to the shorter grazing cycle leading to more immature plants being grazed. Although pastures were strip grazed to maximize efficiency with a high stocking density, the near doubled stocking rate in 2016 was not advantageous in ADG or gain per acre.

[1]Forage Variety Testing Manager, Mississippi Agricultural & Forestry Experiment Station, Mississippi State University (Joshua.White@msstate.edu); Extension Forage Specialist, Center for Forage Management & Environmental Stewardship, Mississippi State University (Rocky.Lemus@msstate.edu); Associate Research/Extension Professor, South Mississippi Branch Experiment Station, Mississippi State University (Daniel.Rivera@msstate.edu).

LOW-INPUT, FORAGE-BASED HAIR SHEEP PRODUCTION

Stephan Wildeus[1]

Hair sheep are well suited for pasture-based production systems due to their ability to utilize moderate quality forage, cope with gastrointestinal parasitism, and lamb on pasture. Virginia State University has managed a flock of landrace hair sheep (Barbados Blackbelly and St. Croix) under a forage-based system for many years, and transitioned towards year-round grazing by integrating warm and cool season annuals, and stockpiled fescue. The flock is managed under accelerated mating with ewes bred at 8-months intervals (November, July and March) and lambing on pasture. Weaned lambs are transitioned to a rotational grazing system with novel endophyte fescue and bermudagrass pastures, and their growth performance and parasite burden is monitored. The use of terminal sire mating and agro-byproduct supplementation was also evaluated to improve productivity. In this system, ewes generally achieve pregnancy rates of 95% (November and July mating) to 75% (March mating), and wean 80 to 95% of lambs born, depending on season of lambing. Terminal sire mating (Dorset rams) reduced pregnancy rates following March mating, but increased birth and weaning weights by 15 to 30%, with no effect on lamb survival. Purebred landrace lambs under rotational grazing achieved growth rates ranging from 0.11 to 0.26 lb/day, depending on seasonal pasture quality. Crossbreeding increased growth rate by 25 to 75% and soy hull supplementation (at 2% of body weight) by 75 to 175%, respectively, again dependent on season of year. Purebred lambs required no deworming, and can provide marketing opportunities for naturally raised product.

[1]Agricultural Research Station, Virginia State University, Petersburg, VA 23806 (swildeus@vsu.edu).

COMPARISON OF ALMANAC AND APSIM FOR SIMULATING OLD WORLD BLUESTEM GROWTH AND WATER USE

Y. Xiong, C. West[1]

Under the severe situation caused by the recent drought and decreasing Ogallala Aquifer, accurate applications of agriculture simulation models will benefit both researchers and producers in the Texas High Plains to investigate agricultural practices with higher water sustainability while maintaining economic productivity. This research was conducted to compare two agriculture model framework: The Agriculture Land Management Alternative with Numerical Assessment Criteria (ALMANAC) model and the Agriculture Production Systems sIMulator (APSIM) model on WW-B.Dahl Old World bluestem [*Bothriochla bladii* (Retz) S.T. Blake] (Bdahl) pasture near Lubbock, TX. Bdahl is an introduced, highly water-efficient and grazing-tolerant perennial C4 grass. Both ALAMANC and APSIM models are capable of simulating plant growth and biomass productivity via plant physiological and plant-soil hydrological parameters using weather, management and plant information. Even though the two models have been developed and enhanced with forage species through decades, there is no validation on Bdahl. The current research aims to refine the models adaptation on limited input Bdahl pasture incorporating grazing effects.

[1]Research assistant, Plant and Soil Science, Texas Tech University (victoria.xiong@ttu.edu); Professor and Thornton distinguished chair, Plant and Soil Science, Texas Tech University (chuck.west@ttu.edu)

AUTHOR INDEX

AUTHOR INDEX